MAKING EDEN

*For Juliette and Joshua*
*and*
*My parents and their insistence on the holiday diaries*

how **plants** transformed a **barren planet**

DAVID BEERLING

OXFORD
UNIVERSITY PRESS

OXFORD
UNIVERSITY PRESS

Great Clarendon Street, Oxford, OX2 6DP,
United Kingdom

Oxford University Press is a department of the University of Oxford.
It furthers the University's objective of excellence in research, scholarship,
and education by publishing worldwide. Oxford is a registered trade mark of
Oxford University Press in the UK and in certain other countries



First Edition published in 2019
Impression: 1



Published in the United States of America by Oxford University Press
198 Madison Avenue, New York, NY 10016, United States of America

British Library Cataloguing in Publication Data
Data available

Library of Congress Control Number: 2018965379

ISBN 978–0–19–879830–9

Printed and bound in Great Britain by
Clays Ltd, Elcograf S.p.A.

**botany** /ˈbot(ə)ni/ n. 1. the study of the physiology, structure, genetics, ecology, distribution, classification, and economic importance of plants. 2. The plant life of a particular area or time. **botanic** /bəˈtanɪk/ adj. **botanical** /bəˈtanɪk(ə)l/ adj. **botanically** /bəˈtanɪk(ə)li/ adv. **botanist** n. [botanic via French *botanique* or Late Latin *botanicus* from Greek *botanikos,* from *botanē* 'plant': botany is from botanic]

*OED*

# PREFACE

My ancestors farmed the south-east corner of England for three centuries, tending to sheep on the lowland pastures of the Kent marshes, reclaimed from the clutches of the English Channel by the Romans and later the Saxons. Another branch of the family has been cultivating pasture and dairy cattle on the farmlands of Essex since before the last world war. The poet John Betjeman (1906–1984), a fine chronicler of life in England (whose statue, incidentally, can be found inside St Pancras Station), wrote of the vanished beauty of nearby Romney Marsh 'where the roads wind like streams through pasture, and the sky is always three-quarters of the landscape'.

For the farming community, life was harder and less romantic than Betjeman's take on things. Shepherds tended large flocks of sheep that grazed hundreds of acres of pasture. Shearing, lambing, rescuing animals from drainage ditches, building pens, and treating wounds were routine matters for them. They holed up at night in simple huts, each a basecamp for storing tools and medicines that offered few comforts and afforded minimal protection from the severe winds that raced in unchecked from the sea. Peripatetic shepherding—one man, his dog, a hut, and large flocks of sheep—was a hard way to make a living and produced a tough breed of men. Their lives depended on a collective wisdom, knowledge handed down from one generation to the next, that ensured stocking densities were adjusted and flocks moved to the best pastures available as the seasons turned.

The shepherd's livelihood depended on getting things right for the landowner by understanding how pastures fattened the white woolly grazing herbivores. The payoff came when the fattened animals were sent to the London meat market and the wool sold to the Wealden cloth industry or smuggled to the European mainland. Wool was a valuable commodity in those days, and the marsh men were quite prepared to risk a jail sentence for the monetary rewards to be had from smuggling and selling fleeces abroad, especially to France. Romney Marsh sheep keepers maintained the pastoral economy in this way for over 500 years and surviving huts can still be found dotted across the region, relicts of times past.

My grandfather's tools hang up in the shed I can see from my office window and connect the past with the present. The handles worn smooth by over five decades of hard labour on the marshes remind me of his understanding between plants, stock, and the land. He understood that the grazing pastures of Kent fuel the rural economy of the county and have done so for centuries, and this regional situation is really a parable for how the whole planet operates. Plants are the food stock of the biosphere and everything else follows from there. Without plants, there would be no us.

Over 7 billion people (and rising) depend on plants for healthy, productive, secure lives, but few of us stop to consider the origin of the plant kingdom that turned the world green and made our lives possible. The origin of humans is frequently considered in books and TV documentaries, but not the origin of land plants. And as the human population continues to escalate, our survival depends on how we treat the plant kingdom and the soils that sustain it. The evolutionary history of our land floras, the story of how plant life conquered the continents to dominate the planet, is fundamental to our own existence.

This, then, is the subject of *Making Eden*. Building on the foundations established by generations of scientists, it reveals the hidden history of Earth's sun-shot greenery and considers its future prospects as we farm the planet to feed the world. Our evolutionary journey stretches back over half a billion years with twists and turns decoded from clues encrypted in fossils, DNA molecules, and the ecology of the plant kingdom. It is a story of how plant life on land originated from freshwater algal ancestors, how it inhaled, diversified, and spread out to conquer the continents, slowly air-conditioning the planet. Finally we are glimpsing answers to the question of origins that have haunted botanists ever since Darwin.

In some respects, *Making Eden* can be regarded as the prequel to my previous book, *The Emerald Planet* (2007), which actually had rather little to say about how plant life on land got going and sustains the diversity of life there. Instead, it offered a closely argued case for recognizing plants as a 'geological force of nature', drivers of changes, sculpting continents, and changing the chemistry of the atmosphere and oceans, to set the agenda for life on Earth. It was a long overlooked story that went on to sow the seeds for the three-part BBC2 television series *How to Grow a Planet* (2012).

But now the stage is set for *Making Eden*. My hope in writing this book is that it might at least give readers pause for thought before dismissing botany as boring and irrelevant. I also hope that it might persuade readers to think of plants, and the scientists who study them, in an entertaining new light. Steve Jones, a leading geneticist at University College London, and one time regular science columnist for the *Daily Telegraph*, wrote an article a few years ago entitled 'Where have all the botanists gone, just when we need them?' He rightly pointed out the central role of plants in the challenges facing humanity, and the challenges facing the subject of botany, writing 'why do students find the vegetable world so boring when without it we would perish?' Plants and botanists are in need of greater advocacy. Ultimately, I hope readers may come to appreciate that botany is an astonishing and deeply engaging field of scientific enquiry, with immediacy to all life on Earth.

**D.B.**
*Sheffield*, 2018

# ACKNOWLEDGEMENTS

I am fortunate to be located in an outstanding academic department at the University of Sheffield, with many talented and generous colleagues. I offer sincere thanks to Jonathan Leake, Charles Wellman, Sir David Read, Colin Osborne, Ben Hatchwell, David Edwards, Jon Slate, and Pascal Antoine Christin, who all read and commented on various drafts, as well as shared and discussed ideas with me. The department houses a 'writer in residence' scheme funded by the Royal Society of Literature: conversations with successive incumbents, Fiona Shaw and Frances Byrnes, about writing and readers proved enlightening.

I offer warm thanks to the following colleagues from other institutes who kindly took the time to read and comment on early and late-breaking drafts of chapters, or have discussed issues that arose: Phil Donoghue, Burkhard Becker, Yves Van de Peer, Doug Soltis, David Hibbett, Steve Banwart, Chuck Delwiche, Jane Langdale, Nick Harberd, Jill Harrison, Lawren Sack, Peter Franks, Alistair Hetherington, Dominique Bergmann, Joe Berry, Ralf Reski, John Bowman, Stefan Rensing, Kevin Newsham, Chris Berry, Bill Stein, Linda VanAller Hernick, Frank Mannolini, Martin Bidartondo, Jeff Duckett, Dana Royer, Christine Strullu-Derrien, and Jim Hansen. All of these individuals helped sharpen my thinking.

John Bowman and Stefan Rensing kindly shared copies of their unpublished manuscripts on the genomes of *Marchantia* and *Chara*, respectively. David Malloch helpfully discussed his mycological thinking with me, and both he and David Hawksworth provided the back-story on Kris Pirozynski and a picture or two. Peter Raven generously agreed to be interviewed and treated me to a memorable lunch in St Louis made unforgettable by his compelling narrative on the state of the planet, and plant life in particular. Paul Kenrick at the Natural History Museum, London, kindly provided a valuable critique on a late, nearly complete draft of the manuscript and caught some crucial errors. I thank all of these people whose work has helped shape and improve the text. Of course, any errors are my own responsibility.

We all need great mentors, no matter which walk of life we choose. In having the late William (Bill) Chaloner FRS (22 November 1928–13 October 2016) as a mentor for nearly two decades, I was fortunate to have had one of the best. Bill died before I completed this book, but had already offered me his encouraging, critical comments on several chapters. This book is dedicated to Bill's memory; a man who really did 'leave an afterglow of smiles when life is done'. In 1970, Bill published an influential academic paper entitled 'The rise of the first land plants' (*Biological Reviews*, **45**, 353–77) in which he dealt with 'the facts from which a hypothesis of evolutionary progression may be constructed'. Nearly 50 years on, I realize that I have unconsciously written my own update on that same thesis.

Financial support for my research group over the past decade has been provided by the Royal Society, the Leverhulme Trust, and the Natural Environment Research Council, UK, for which I am most grateful.

Sincere thanks must go to my patient and enthusiastic editor, Latha Menon, whose editorial comments and wise suggestions improved the text. My gratitude too, to Jenny Nugée, and the rest of the Oxford University Press team, who efficiently took the book through the publication process.

Writing this book has taken quite some time, which in one way has been helpful because the suspicion is that a long manuscript improves with a long gestation time. But the down side is the steeply rising forbearance required of my wife, Juliette, who additionally provided title inspiration. So there comes a point, after so many years, when you have to get on with it. Joshua's imminent arrival helped daddy get a move on, so that we can all spend more time together growing apples, walking, bird watching, and learning how to fish. I thank Juliette and Joshua for their patience, love, and support, and look forward to our fun times together now that this is done.

# CONTENTS

# 1 ALL FLESH IS GRASS

'Going up that river was like travelling back to the earliest beginnings of the world, when vegetation rioted on the earth and the big trees were kings.'

Joseph Conrad, *Heart of Darkness*, 1899

The spectacular rise and diversification of plant life on land reshaped the global environment, and the possibilities of our lives. It was one of the greatest revolutions in the history of life on Earth. But there's no need to take my word for it. Here is palaeontologist Richard Fortey's view on the matter, writing in his book *Life: An Unauthorised Biography*, 'There cannot be a more important event than the greening of the world, for it prepared the way for everything that happened on land thereafter in the evolutionary theatre'. This book tells the story of how it happened. Or, at least, the story of how we think it may have happened. A celebration of discovery and scientific enquiry, this book lifts the lid on the evolutionary story of how plants won the land. How the Earth went from being a dull, rocky, naked planet to today's world cloaked in a wonderful diversity of plant life on which we all depend for our very existence. It looks also to the future as the sixth great extinction in the history of life on Earth looms unwantedly and alarmingly on the horizon.

First, though, let us take a step back. Imagine, for a moment, a strange alternative world without plants: a naked Earth shorn of its greenery. This alien planet is a favourite haunt of fiction writers and directors of post-apocalyptic movies, and for good reason. In a world where plant life never evolved to cloak the continents in green, never changed the fabric of the landscape or the cycling of elements through the biosphere, Earth's planetary prospects for life support look bleak. Its barren, windswept landscapes are coloured in drab mackerel greys and monkey browns; leafless, treeless, grassless, and useless for supporting a diversity of animal life.

When science fiction writers describe a vision of humanity's dystopian future, it is no accident that the destruction of plant life is a key motif. In John Christopher's haunting sci-fi classic, *The Death of Grass*,[1] a fictional virus wipes out the crops that feed humanity, and the grasses that feed the cattle that feed the people. Mass starvation decimates Asia and when the deadly virus hits Britain, beleaguered survivors quickly discover that, when there is nothing to eat, society degenerates with alarming speed. In J.G. Ballard's vividly imagined *The Drought*,[2] industrial waste has produced a mantle of artificial polymers over the oceans, destroying the hydrological cycle and transforming the planet into a wilderness of dust and fire. Ballard's survivors 'follow the road upwards, winding past burnt-out orchards and groves of brittle trees like the remnants of a petrified forest'. Half a century later, Cormac McCarthy's *The Road*[3] depicts a ravaged landscape with charred dead trees, rooted in scorched earth and coated in silver by drifts of ash. McCarthy introduces his survivors into this post-apocalyptic landscape where there is nothing to eat, no plants and no cattle; civilization has crumbled and, like the protagonists of Christopher's novel, they face the terrifying dangers of a degenerate society. Grasses, incidentally, have form for bending animals to their collective wills and John Christopher also succumbed subconsciously to their charms. He settled in the pleasant medieval coastal town of Rye in East Sussex, the only town in England named after a grass.

Fictional works by Christopher, Ballard, McCarthy, and other practitioners of the post-apocalyptic genre succeed partly because they recognize that a world without plants spells disaster for humanity. This theme resonates with our own deeply rooted concerns about food and survival. Movie directors too have tapped into these anxieties. In Christopher Nolan's 2014 blockbuster *Interstellar*, repeated crop failures slowly render Earth uninhabitable, prompting the need to evacuate the population to a new planetary home via a wormhole. The following year, Ridley Scott's box office hit *The Martian* saw Matt Damon play astronaut Mark Watney marooned on Mars. Watney's immediate concern is with the need to grow food. Being probably the only astronaut who ever trained as a botanist, he naturally harnesses his skills and expertise to improvise a potato garden, utilizing Martian soil fertilized with human waste.

The Romans recognized the central importance of plants to our world too. Any aspect of life deemed important enough had its own dedicated god or goddess, and Ceres was their goddess of agriculture, depicted on Roman coins with a wheat crown standing on a chariot drawn by winged serpents. They worshipped Cer because without her blessings, harvests might fail and starve the Empire. Our modern tribute to Ceres has been to name the largest asteroid in the inner Solar System after her. Ceres sits in an asteroid belt between Mars and Jupiter.[4] Remote sensing surveys[5] suggest Ceres (and Mars) contain deposits of the same sea-floor minerals that perhaps sparked life on a young Earth, billions of years before plants made land. Current ideas for the origin of life favour deep-water hydrothermal vent settings, with the microbes involved drawing their energy from seawater chemistry rather than from the Sun.[6] 'White smoker' hydrothermal systems are currently prime candidate locations. Forming when minerals deep in the fractured oceanic crust react with seawater, white smokers are distinct from the hot, acidic 'black smokers' that give birth to new sea-floors as the continents shift apart.[7] The warm (ca. 70°C), alkaline hydrothermal fluids of white smokers move up through the splintered crust to emerge at the sea floor rich in dissolved hydrogen.[8] Towers of calcium carbonate develop, reaching upwards 60 metres or more from the sea floor, each riddled with networks of tiny pores and adorned with feathery fans of minerals. These porous spires, bathed in volcanic hydrogen-rich effluent, may have offered the 'goldilocks' environment for booting up life—not too hot, not too cold, and not too acidic.[9]

Today's green continents are an evolutionary legacy of those distant events marking the dawn of life. Instead of drawing energy from seawater chemistry like their microbial progenitors, plants harvest the solar energy showering our planet. That free source of energy has travelled an astonishing 93 million miles in just eight minutes. Two-thirds of it hits the world's oceans, where it drives photosynthesis by marine plants, mainly free-living phytoplankton. These microscopic plants form the base of the oceans' food chains.[10] One-third of it hits the land surface, where the leaves of forests, grasslands, and crops capture it to power photosynthesis and synthesize biomass—organic matter—from carbon dioxide and water. By converting solar energy into chemical energy stored in the organic carbon compounds that make up their bodies—the tissues of leaves, roots, shoots, flowers, and grains—plants act as nature's wonderful green energy transducers. Herbivores eat plants, and carnivores eat herbivores, with each group of organisms extracting energy as they inexorably convert plants into flesh. Finally, fungi and bacteria, the microbial heroes of decay, employ a remarkable repertoire of metabolic tricks to feast on the decaying plants and animals, recovering the last remaining vestiges of energy. Plants are the crucial green engines doing all the work to supply the energy that supports life on land. After they convert the energy radiated from the Sun into energy-rich organic matter, no other group of organisms adds energy into the food chain, and everybody else extracts it. It is a straightforward rule of the natural world, placing plants indispensably at the base of the food chain. Without plants, the herbivores starve; without herbivores, the carnivores starve. Without plants, the food chain collapses and there is nothing to sustain terrestrial life. No mammals, no primates, no us. Herein lays the truth of that gnomic saying in the Old Testament Book of Isaiah, 'all flesh is grass'. Our improbable botanist-astronaut Mark Watney understood the point well enough, as did John Christopher in *The Death of Grass*. Christopher's protagonists quickly realize that a virus wiping out wheat, oats, barley, and rye, then supplies of meat, dairy food, and poultry, mean 'all that's left are the fish in the sea'. Satellite observations indicate that the photosynthetic productivity of phytoplankton in the oceans that supports the fishes is roughly equal to that of plants on land. But the productivity of the land is squeezed and concentrated into one-third of the planet's surface, and this explains why most species on Earth (85–95% of all organisms other than microbes) live on land.[11] In forests and grasslands, for example, the species diversity outnumbers that in the oceans by 25 to 1.

The establishment of plant life on land is, then, a prerequisite for sustaining a consumer society of land-dwelling animals. Insects were amongst the first animal groups to stake out habitats on land, feeding on algal mats and early land plants right from the start.[12] As the first-comers, they won the rights to digesting the cellulose that makes up plant remains by using bacteria in their guts, as insects do today, and presumably acquired those bugs by feeding on decaying vegetable matter. When the earliest limbed vertebrates (animals with backbones) crawled ashore during the fin-to-limb transition, they struggled for millions of years to leave their aquatic ways behind. Early vertebrates, the ancestors of modern reptiles, birds, and mammals, inherited air-breathing lungs from their air-gulping lobe-finned fishy relatives. They heaved themselves on to land around 370 million years ago, roughly a hundred million years after plants.[13] Yet these lumbering reptilians were unable to eat plants directly. Encumbered with fish-like mouths, they fed instead by going into the seas to hunt animals. In the water, they could grab food items floating in front of them and use suction to draw it into their mouths and gulp it down. These primordial vertebrates took another 80 million years to evolve jaws and teeth, adaptations that enabled them to properly exploit terrestrial vegetation as a foodstuff.

## 'ANIMALS ARE INFERIOR TO GREEN PLANTS'

Professor Albert C. Seward (1863–1941), Master of Downing College and Vice-Chancellor of Cambridge University, once wrote a no-nonsense book published in 1932 and titled, *Plants: What They Are and What They Do*. Although written some time after the publication of his report on the fossil plants collected by Scott of the Antarctic's ill-fated 1910 *Terra Nova* expedition,[14] this slim volume with its dull green cover failed to become a bestseller. Chapter Three, entitled 'The Superiority of Green Plants to Animals', gives no quarter to the sensibilities of his zoological colleagues. In it, Seward forcibly argues the irrefutable point that plant life is the foundation stone of all living things. He emphasizes the obvious but often overlooked point that animals are unable to use carbon dioxide directly from the atmosphere. They cannot transduce solar energy into chemical energy. Animals obtain energy by burning carbohydrates during respiration that plants manufactured from the sugars synthesized by photosynthesis. No animal can manufacture its own carbohydrates directly from sunlight, which is why Seward declares that 'animals are inferior to green plants'.

Animals are inferior to plants in other ways too. Few animals live to reach their hundredth birthday, for instance, while the lifetimes of trees usually exceed that milestone. Bristlecone pines enduring the inhospitable heat and dust of the dry mountains of south-western North America live for thousands of years. In the White Mountains of eastern California, the oldest specimen of all is an astonishing tree nicknamed 'Methuselah', after the longest-lived person in the Bible. Estimated to be over 4700 years old, Methuselah achieves its exceptional longevity by repeatedly forming new structures and organs, such as needles and roots. The continuous renewal of essential organs in this way constitutes the tree's winning strategy in the game of life, and plants adopted this modular growth habit from their earliest days on land, half a billion years ago.

Appreciating the essential role plants play in sustaining Earth's astonishingly rich biological diversity begins to make clear why the origin and diversification of plant life on land was such a pivotal chapter in the history of our planet. What could be more important? It paved the way for the evolution of terrestrial animals, and ultimately led to the appearance of human beings, the most complex organisms in the known Universe. Yet the inconvenient truth is that many of us fail to appreciate the essential role plants play in our lives. We are, it seems, suffering from the affliction of 'plant blindness' (PB).[15] Defined as an 'inability to see or notice plants, leading to the inability to recognize the importance of plants in the biosphere and in human affairs', PB is an 'anthropocentric ranking of plants as inferior to animals'.[16]

Although it sounds like something Seward might have come up with, PB was actually put forward by North American academics in the 1990s. At one level the explanation may come down to a question of timescales. The lives of plants unfold on a different timescale to our own. Compared to our lives the actions of plants often seem imperceptibly slow, but it may also be more deep-seated than this.[17] Our ancestral brains form part of a visual monitoring system wired by evolution over millions of years to detect animals rather than plants.[18] Streams of images containing 10 million bits of information are received and transmitted by the retinas of our eyes every second and require visual processing by the optic lobe. From this data onslaught, the brain extracts a mere 40 bits, fully processing only 16 of these to reach our conscious attention. How does the brain decide which crucial 16 bits of information to focus on? The answer, shaped by evolution, is a matter of priority for survival. It searches for movement, colours, patterns, and objects that are potential threats. Our brains are hardwired to be more vigilant at noticing animals than plants and for good reason. You could imagine a plausible evolutionary scenario in which human survival rests on identifying fellow humans as possible mates or foes, and recognizing animals not plants as a deadly threat or a potential meal. Few plants are as immediately life-threatening and urgently demand our attention in the same way as snarling, slavering predatory animals. Even John Wyndham's fictional triffids are not fast, agile, life-threatening green aliens. Instead, he invents them as lumbering opportunists, farmed and domesticated for their valuable vegetable oil, with stings docked, until their time comes. Wyndham's novel may lack psychological depth but is, in part, a Darwinian parable in which an environmental catastrophe threatens human survival, as a strange meteorite shower blinds all those who observe it. In this literal manifestation of PB, plants exact the ultimate revenge; incapacitated, sightless humans preyed upon by deadly, superbly adapted, ambulatory, carrion-eating triffids.

Michael Crichton, through the eyes of palaeobotanist Ellie Sattler in *Jurassic Park* (1993), persuasively sums up the issue like this:

> people who imagine that life on earth consisted of animals moving against a green background seriously misunderstood what they were seeing. That green background was busily alive. Plants grew, moved, twisted, and turned, fighting for the sun; and they interacted continuously with animals – discouraging some with bark and thorns, poisoning others; and feeding still others to advance their own reproduction, to spread their pollen and seeds. It was a complex, dynamic process which she never ceased to find fascinating. And which we she knew most people would never understand.

No wonder it is not easy seeing green and shaking off our deep-rooted zoo-chauvinism. Our brains are wired to notice animals, not plants, and our innate PB casts a new light of forgiveness on editors running on zoo-centric software. In 2009, the international science journal *Nature* reported '15 evolutionary gems' as part of the celebration of the bicentennial anniversary of Charles Darwin's birth since the publication of the *Origin of Species*, intended to give readers 'a resource…for those wishing to spread awareness of evidence for evolution by natural selection'.[19] In an extraordinary oversight, plant life is conspicuously absent from the list that features, amongst other organisms, snakes, guppies, fruit flies, whales, lizards, birds, and water fleas. The top three gems from the fossil record were: #1. the land-living ancestor of whales, #2. from water to land, and #3. the origin of feathers. The explanatory text for #2 immediately captures our attention. Here we are told that 'the *first* transition from water to land took place more than 360 million years ago', and that 'it was one of the most demanding such moves ever made in the history of life. How did fins become legs? And how did the transitional creatures cope with the formidable demands of land life, from a desiccating environment to the crushing burden of gravity?' Plants made land, and evolved remarkable adaptations to solve the tricky challenges of making a living out of thin air, a hundred million years before vertebrates.

We might look to Darwin's *The Origin of Species* for some answers to questions concerning the mysterious origins of terrestrial plant life. Yet after scrutiny of all

460 dense pages of the evolutionary master's elegant prose, we finally discover that his magnum opus remains silent on the thorny issue of the origins of land plants. Darwin was no fool. He builds a highly original picture of the struggle for existence and evolution by natural selection through Herbert Spencer's (1820–1903) memorable phrase 'survival of the fittest'. His theory of evolution was an incendiary intervention in the midst of Creationist orthodoxy, but he refused to be drawn into publishing his thoughts on the question of the origin of life, preferring instead to confine them to a private letter to his friend the botanist and Director of the Royal Botanic Gardens, Kew, Sir Joseph Hooker (1817–1911). Here he cautiously asks 'if (and oh what a big if) simple life may have got going "in some warm little pond"...'. He showed similar reticence when it came to the evolutionary origin of land plants, for he wrote little about it in Chapter X, 'On the Imperfection of the Geological Record' or Chapter XI, 'On the Geological Succession of Organic Beings'.

Ultimately, the challenge of explaining the mystery of how the barren rocky continents acquired their cloaks of green awaited Frederick Bower (1855–1948). Long-time occupant of the Chair of Botany at the University of Glasgow, Bower stands out as a giant in the history of botany.[20] He studied botany at Trinity College, Cambridge as an undergraduate but was unimpressed by the teaching, describing it as 'moribund in summer and actually dead in winter'. Nevertheless, he graduated with first-class honours, and as a young man in his twenties he was fortunate to have witnessed the University of Cambridge awarding Darwin an honorary degree.[21] Who knows what inspiration the young Bower drew from this experience? What we do know is that he went on to be an inspiring lecturer, imbuing his students and colleagues with an enthusiasm for botany during its heyday in the Victorian era. We also know that his own research began just as Darwin's findings from the *Origin of Species* were becoming effective in biology. His great hypothesis for how land plants arose to conquer the terrestrial realm is set out in *The Origin of a Land Flora* (1908), its title a homage to Darwin's great book. Bower recognized 'that certain algae represent in their general characters the original source from which the land flora sprang'. Although he had no idea which group of algae gave rise to land plants, Bower was right. He was also right with his conjectures about the evolutionary dodge required for successful reproduction on the land. And yet, as befits our concerns with PB, few historical treatments of modern science mention Bower. John Gribbin's richly entertaining *Science: A*

*History*,[22] for example, overlooks Bower and Seward and, for that matter, the great botanical explorer Sir Joseph Banks (1743–1820), holder of the prestigious position of President of the Royal Society, London with distinction for over 40 years.

Ever since Bower's time, if not before, scientists have been minded to persuade Earth's floras to tell secret tales of their remote past. Insights and discoveries from the frontiers of modern science over the past few decades have melded with the great leaps and incremental steps taken by earlier generations of scientists. Rich new sources of knowledge and understanding of a vanished ecology are offering answers to explain how a world without plants turned green. Our narrative of the evolution of plant life on land stretches back over the past 500 million years. It is framed by the family tree that portrays the evolutionary relationships of the plant kingdom (Chapter Two). All land plants, we discover, arose from an ancestral freshwater algal lineage that colonized the land only once—an evolutionary singularity that changed the world, forever. Over the next 500 million years, there followed an unstoppable succession of green evolution, as ferns, forests, and grasslands appeared and diversified. The success of sun-shot ferns and forests testifies to the 'pull of the land' as they exploited the ecological opportunities created by making a living out of thin air. The greening of the land was an inevitable outcome of the plant kingdom's sharply escalating power as it began annexing more and more land.

Questions relating to 'how' the evolutionary action portrayed by the evolutionary tree might have taken place are addressed in two chapters reading the genetic code of plant life (Chapters Three and Four). Chapter Three shows how recovering molecular memories stored in the genomes of living plants offers a new and highly original perspective on their evolutionary insistence and global success. Written into the genomes of photosynthesizers—from ubiquitous algae to towering conifers—are clues to a shifting genetic make-up at critical moments in their evolutionary history. Often these coincide with critical moments in Earth's history, and for good reason, because the stories of plants and the Earth's are entwined. Decoding the genomes of our modern floras is telling us new stories from the past about how the kingdom of plants diversified and survived the mass extinctions that befell the animal kingdom. Indeed, the extraordinary suggestion that genomes of many living plants duplicated millions of years ago points to a genetic impetus for the explosion of biodiversity that followed the origins of seed plants and flowering plants .

Duplicating entire genomes is one thing. Wiring up and creatively recycling old bits of duplicated DNA, some of it inherited from freshwater green algae, into genetic toolkits for building land plants is quite another. The novelist J.G. Ballard anticipated the discoveries described here in Chapter Four. In a strange short story called 'Prisoner from the Coral Deep'[23] Ballard writes of a school teacher walking a deserted stretch of the Dorset coast. On finding a coiled, fossilized shell, the teacher asks 'if only one could unwind this spiral, it would probably play back to us a picture of all the landscapes it's ever seen—the great oceans of the Carboniferous and the warm shallow seas of the Triassic'. Now, half a century later, Ballard's surrealist thought experiment is becoming a scientific reality. As Chapter Four reveals, we are now unwinding the coiled DNA strand inside the nucleus of living cells to shed light on evolutionary events long ago. These fascinating glimpses into the past come from evolutionary developmental biology ('evo-devo'), a branch of science uniting genetic processes within the nucleus with the development of cells, tissues, and whole plants. Plants, we discover, evolved by modifying core genetic toolkits with many components inherited from algae. Innovations such as roots and leaves, and essential chemical messenger systems informing the plant when and where to grow, and in what direction, were made possible through expansion and repeated rewiring of different toolkits.

Although roots and leaves are often regarded as standout evolutionary accomplishments, ultimately they had to be combined with microscopic gas valves—stomata—to build successful land plants. Stomata proved to be a secret of the plant kingdom's success on land (Chapter Five). These tiny mouths appear in the fossil record millions of years before roots or leaves. Exquisitely tuned to the environment, stomata adorned the delicate shoots of early land plants as they poked their heads above primeval soils into the drying sunshine. They allowed plants to control the loss of water evaporating from cells and tissues as carbon dioxide moved in the opposite direction, fuelling photosynthesis. Their appearance marked the beginning of a radical new mode of plant life based on exploiting water and nutrients obtained from the soil and enabled it to expand into places it had never gone before. Once plants evolved the 'trick' of making stomata, Earth's floras breathed deeply, grew taller and spread outwards across the continents. Plants evolved no better solution for swapping carbon dioxide for water and, over 400 million years later, modern plants still rely on stomata packed onto their leaves to breathe, made with the same genetic machinery that originated in their distant ancestors.

Yet the triumph of the plant kingdom in dominating the continents involved more than the development of innovative cells and tissues courtesy of a flexible repertoire of genetic toolkits. The move to land required an often-overlooked innovation: an enduring partnership between plants and symbiotic fungi. Coined in 1877 by the German biologist A.B. Frank (1839–1900), the term symbiosis means 'two species that live together'. Frank also came up with the term mycorrhiza for the association, from the Greek for fungus (*mykos*) and root (*riza*).[24] Sadly, he did not live long enough to see his concept of mycorrhizal symbiosis gain a revolutionary new perspective with the discovery of 410-million-year-old fossilized fungal threads penetrating rootlets of early land plants.[25] What these and other fossils don't tell us—what they cannot tell us—is how ancient alliances between plants and fungi may have operated to facilitate the emergence of terrestrial plant life (Chapter Six). Answering these questions takes us on a botanical expedition to New Zealand's South Island, where an ancient lineage of liverworts dwells in the moss-draped old-growth southern beech forests. Liverworts are simple plants, surviving relics from a forgotten world, and close living relatives of the earliest land plant pioneers. Fossils, DNA, and experiments with these odd-looking photosynthesizers are building a compelling new picture of how soil fungi helped plants gain the land. Plants and fungi are locked in a tight symbiotic alliance, dating to the origin of the terrestrial biosphere, that underpins the health of forests, grasslands, and croplands worldwide.

But we should not imagine for a moment that the great greening of the land occurred in isolation from the rest of the planet—far from it. The dramatic foresting of the landscape had consequences; the complex interdependence of energy and matter demands it. It had consequences for the rocks into which trees sank their roots, for the chemistry of the atmosphere, for the chemistry of the oceans, and for all life. It took decades to unpick what happened but diverse lines of evidence are building a new world view recognizing trees and symbiotic fungi as the bioengineers of global change over millions of years. By destroying rocks and minerals, the roots and networks of their 'rock-eating' fungal partners crumble Earth's rocky crust by breaking it down into soil, capturing the greenhouse gas carbon dioxide in the process and storing it for millennia in the oceans. Chapter Seven explains how the transformation of the naked planet to a forested Eden began sculpting climate and creating rich, nutritive soils to make Earth habitable. As we shall discover, the secret lives of plants hold great relevance today, guiding us towards strategies for mitigating climate change by capturing and storing carbon dioxide emitted by our burning of fossil fuels for energy.

There is but one way to end this look at the profound transformation of the planet by the greening of the land: by considering Eden's future in the face of escalating threats from humanity (Chapter Eight). What does the future hold for the great diversity of plants that make up the spectacular floras of the world? Humanity's future hinges on how we treat the extraordinary green legacy of those early land dwellers that adapted to life out of water half a billion years ago. Earth's biodiversity is declining rapidly as we unsustainably exploit natural resources and destroy natural ecosystems to meet the rising food and energy demands of 7 billion people. Despite the considerable efforts of Peter H. Raven, the former director of the Missouri Botanical Garden, described by *Time* magazine as a 'Hero for the Planet', all the metrics agree. Our 'ecological footprint' has never been larger, and the threat to nature is escalating as the human population rises. Safeguarding the future of the planet, as I'll argue in Chapter Eight, requires mitigation and adaptation as the only sensible responses[26] if we are to avoid the sixth great wave of extinction in the history of life on Earth.

The arch-Darwinian evolutionary biologist Richard Dawkins rightly claims that evolution is '*The Greatest Show on Earth*'.[27] With characteristic flair, he writes 'short of rocketing into space, it is hard to imagine a bolder or more life-changing step than leaving water for dry land'. Dawkins is actually writing about the evolutionary appearance of the first land animals, but never mind; he makes a good point. It is indeed harder to imagine a bolder and more life-changing step than moving from water to dry land. To be struck by the full force of what evolution can achieve, there is no better group of organisms to look at than life-sustaining green plants. Not only did plants take the giant step on to dry land 100 million years before four-legged vertebrates, but their diversity *outnumbers that of vertebrates by 10 to 1*. Understanding how plants made the transition from life in freshwater to dominate successfully the terrestrial environment has planetary significance now and in the future. Understanding how a world without plants turned green is one of the big unanswered questions in science, second only to the origin of life itself. In Fortey's words again, it 'was an extraordinary transformation in the beauty of the world, not merely an opportunity for a great expansion in the compass of life'.

We are about to embark on an exciting but challenging journey that offers the ultimate reward—a solution to one of the great mysteries in the history of life on Earth. Chronicled in the exquisite fossil record reaching back hundreds of millions of years of plant life is the story of what happened. Chronicled in the genes and ecology of organisms alive today is the story of how it happened; how the world turned green.

# FIFTY SHADES OF GREEN

'Even a botanist must occasionally take off his rubber boots and log onto a database. Kinships between plants are no longer determined by fiddling with petals or reproductive organs but by comparing genetic sequences among species.'

Lone Frank, *My Beautiful Genome*, 2011

A road trip out of Los Angeles is admittedly not an obvious starting point for a journey pursuing the remote ancestors of our modern land floras. The natural landscape has long since vanished in the city, buried beneath a metropolis sprinkled with palm trees imported to add a refreshing splash of greenery for urban citizens. Continue travelling north-east out of LA, though, and eventually you encounter the stunning natural terrain of western North America with mountain ranges and expansive deserts looming on the horizon. Before long, the Mojave Desert (named after the Mohave tribe of Native Americans) makes its hot, arid presence known as the heat haze shimmers above the parched land. Step outside the comfortable air-conditioned car and a furnace-like dry wall of heat immediately hits you. Under these conditions, it is easy to become distracted and overlook the communities of photosynthetic organisms living beneath your feet. These flattened crusts of life persist, scattered between the Joshua trees (*Yucca brevifolia*) standing out against the cloudless desert sky, clinging to dusty desert rocks and thin soils. Drab and dull though they might seem, these crusts are actually hotbeds of biodiversity bristling with life,[1] despite enduring some of the most extreme and inhospitable environments on Earth.[2]

Cryptobiotic crusts, as they are known, occur throughout the arid regions of the world, and contain cyanobacteria, similar to the earliest terrestrial fossilized forms found in 1.2-billion-year-old rocks, living together with algae and fungi.[3] They offer a fascinating glimpse of life on land long before the arrival of terrestrial plants. Before plants claimed the land, the naked continents were rocky, windswept, and desolate. However, it would be wrong to think of them as totally devoid of life. Photosynthetic communities of cyanobacteria and algae had already arrived, quietly making a living for themselves, for perhaps a billion years or more. Dotting the landscape with the merest hints of green, they foreshadowed what was to come. Through the subtle processes that chemically accelerate the slow destruction of exposed bedrock, these crusts inadvertently prepared the ground for the later spread of plant life. By binding together shattered rock grains with fragmentary particles of organic matter, they created patches of land with the thinnest veneer of consolidated soil, preparing the way for a greener world.

Molecular sequencing of DNA is revealing the hidden diversity of free-living desert green algae in these biological crusts and their close evolutionary connections with freshwater algal ancestors.[4] Occasionally mosses can be found amongst these crusts, rubbing along perfectly well with the fungi, cyanobacteria, and algae. Sand-locked desert mosses charmingly exploit the morning dew that collects on their shoots for reproduction: it provides the liquid medium for their male sperm to wriggle through and fertilize the female egg cells to renew the life cycle. Cyanobacteria, green algae, mosses, and the Joshua trees are related to each other, as Charles Darwin (1809–1882) and Alfred Russel Wallace (1823–1913) brilliantly realized over a century ago. Darwin adopted a tree as an inspired metaphor for conveying the essential point that the great variety of life can be traced back to a single common ancestor, with all branches leading back to a unified trunk. This became the evolutionary tree of life.

The concept of relatedness among all life forms on Earth, past and present, is so familiar to us today that it is easy to forget this has not always been the case. At the close of the seventeenth century, botanists lacked a coherent framework for classifying the diversity of plant life. Sir Isaac Newton (1643–1727) may have begun writing mathematical equations distilling the fundamental physics of the Universe into his grand *Principia Mathematica* around this time, but botanists were still struggling to make sense of the world of plants. The English naturalist John Ray (1627–1703) rode to the rescue when he made the important conceptual breakthrough of classifying plants based on similarities and differences in the physical appearance (morphology) of species. The forgotten man of plant taxonomy, Ray was a contemporary of Newton and rose from being the son of a blacksmith to become a Fellow of Trinity College, Cambridge. His 1686 *Historia Plantarum* established, amongst other things, the modern concept of a species, but it fell to the Swedish botanist and physician Carolus Linnaeus (Carl von Linné; 1707–1778) to establish the science of classification—the systematic ordering of nature. Collecting and studying plants was Linnaeus's first love. He occasionally hit the buffers with his (for the time) controversial principles for classifying flowering plants based on their male and female reproductive parts. His genius found expression, however, when it came to the describing, naming, and grouping of organisms, establishing the principles for classifying them into a hierarchy of relationships. Not a man noted for his modesty, his favourite aphorism was 'God creates and Linnaeus classifies'. He also wrote four autobiographical memoirs, published after his death, to ensure his achievements would outlive him.[5] A charming and popular teacher, Linnaeus was admired and respected by his contemporaries and students alike.

Darwin and Wallace, modest men by comparison, explained how the diversity of life obsessively catalogued by the Swede came about. Their celebrated theory of evolution by natural selection revolutionized biologists' world view as much as Newton's *Principia* had revolutionized physics and astronomy nearly two centuries earlier. At a stroke, they breathed evolutionary life into the dry classification of organisms undertaken by Linnaeus as he organized our knowledge of the living world. Wallace and Darwin gave biology the foundations of a grand unified theory.

Taking their cue from the concepts of relatedness expressed in the tree of life, the task for taxonomists is to assemble a grand picture of the evolutionary relationships that link together all species of the plant kingdom.[6] The tree of plant life currently depicts the patterns of relationships for the three major lineages of plants: glaucophytes (freshwater algae), rhodophytes (red algae), and the Viridiplantae (chlorophytes, charophytes, and land plants).[7] Glaucophytes (also known as glaucocystophytes) are obscure microscopic freshwater algae and not an important group to our story. Red algae and green algae, on the other hand, are both species-rich, diverse groups dwelling successfully in a wide range of environments. Red algae did not give rise to land plants, but nonetheless deserve a mention, not least because the oldest accepted plant fossils, dated to 1.2 billion years ago, are the remains of red algae. Preserved in rocks of the Canadian high Arctic, these early photosynthesizers are called *Bangiomorpha pubescens*, because their whiskery filaments, around 2 mm long, closely resemble the bristling living red alga *Bangia*.

They grew attached to shoreline rocks with holdfast structures.[8] Red algae get their colour from specialized light-harvesting complexes of pigmented proteins called phycobilisomes. The proteins are organized into elegant geometrical arrangements for efficiently capturing those wavelengths of light not filtered out by seawater, providing a crucial source of solar energy for photosynthesis. Similar pigments are used by the glaucophytes and cyanobacteria.

At the base of the tree there sits a mysterious alga that originated in the primeval oceans around two billion years ago, in the far-off days of the Palaeoproterozoic.[9] Destined to shape the future of the planet, and to secure the centrality of plants to life on Earth, it flitted through the ancient oceans powered by a whip-like structure called a flagellum. And it had turned green by acquiring, in an extraordinary way, a chloroplast for conducting the business of photosynthesis.

Chloroplasts are microscopic bodies that are crowded into plant cells. They contain chlorophyll, the pigment that intercepts sunlight and gives them their green colour. The acquisition of chloroplasts marks the origin of plants' ability to photosynthesize; a time when harvests of solar energy and carbon dioxide started fuelling the biosphere, first in the oceans and then on land. Lynn Margulis (1938–2011), an influential and flamboyant character from the world of cell biology, calls this a tale of hostage-taking, slavery, and domestication. It all started in the oceans, when a host cell engulfed a free-living photosynthetic cyanobacterium for food.[10] Instead of being digested, the cyanobacterium survived and took up permanent residence; its fate was to become a chloroplast. Chloroplasts, it turns out, are really photosynthetic bacteria trapped inside plant cells (**Figure 1**). Millions are packed into leaves and stems, each one a miniature photosynthetic factory, or if you prefer, an energy transducer, using its chlorophyll pigments to trap solar energy and convert it into chemical energy for synthesizing organic molecules from carbon dioxide and water. By an astonishing trick of nature yet to be fully understood, chloroplasts are persuaded to hand over the sugars they manufacture by photosynthesis to fuel the host cell's metabolic activities. Why the complex union between an alga and a cyanobacterium took place remains a mystery. We might speculate, as others have, that some change in local environmental conditions, coupled with a scarcity of prey, made it beneficial to stop eating and start photosynthesizing. Regardless of the why, these benevolent bodies also conveniently replicate themselves before a plant cell undergoes the upheaval of dividing into two daughter cells, ensuring that each cell inherits the ability to

## LIVING TOGETHER—THE THEORY OF ENDOSYMBIOSIS

The acquisitions and mergers of the cellular world, explaining the origin of the chloroplast and mitochondrion organelles, are part of a phenomenon called endosymbiosis; meaning literally 'living together inside'. The theory of endosymbiosis championed by Margulis is widely accepted by the broader scientific community and was originally proposed at the dawn of the last century by the Russian biologist Konstantin Mereschkowski (1855–1921).[11] As is often the case when a radical new theory appears, fierce scepticism greeted the arguments, no matter how carefully laid out before colleagues. Had he lived, Mereschkowski would have had the last laugh because his endosymbiosis theory is now part of mainstream biology, having found an advocate in Lynn Margulis.

In her lifetime, Margulis saw the discovery of thousands of genes inherited from the engulfed photosynthetic cyanobacterium in the genomes of green plants. It is worth quoting here from her obituary written by James Lake, of the University of California, Los Angeles, in the leading scientific journal *Nature*.[12] Lake wrote of the moment when Margulis's paradigm-changing view first received serious support: 'As Boston University's Douglas Zook, then an undergraduate in one of her classes, recalled, it was an emotional moment in 1978 when her ideas on endosymbiosis were confirmed. She strode into class beaming, holding Robert Schwartz and Margaret Dayhoff's classic paper, "Origins of prokaryotes, eukaryotes, mitochondria, and chloroplasts",[13] hot off the press. That paper concluded: "The chloroplasts share a recent ancestry with the blue-green algae [cyanobacteria]", and that "the mitochondrion shares a recent ancestry with certain respiring and photosynthetic bacteria, the Rhodospirillaceae". Margulis's proposals for endosymbiotic chloroplast and mitochondrial origins had both been proven in the same paper.' Based on this account, it is fair to say that 1978 marks a suitable date for the entry of the theory of endosymbiosis into mainstream science.

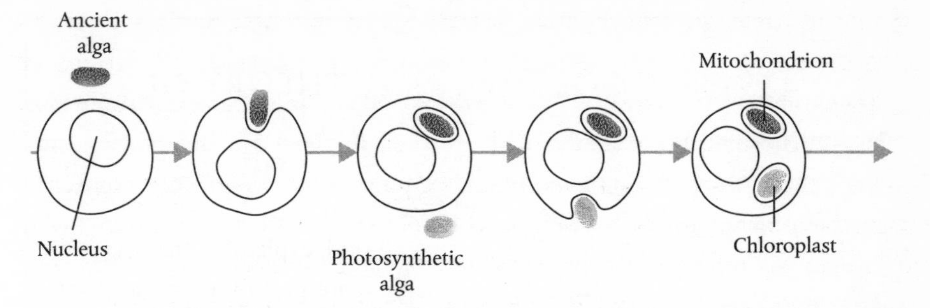


**Figure** 1 Hostage taking and slavery in the ancient oceans. An ancient alga acquired organelles for photosynthesis (chloroplast) and for generating energy by burning oxygen (mitochondrion).

photosynthesize with its own population of chloroplasts.[14] Green plants have been using the same basic photosynthetic machinery, packed into chloroplasts, more or less unchanged for nearly two billion years since they first captured a once free-living bacterium.[15]

Even before the last common ancestor to all green plants irreversibly assimilated a cyanobacterium, it had already pulled a similar trick to gain something else it needed: energy. Small structures inside cells called mitochondria are the vital tiny battery packs generating the chemical energy essential to life. They use elaborate biochemical pathways that burn oxygen and sugars to fuel the cell, a process known as aerobic respiration. Margulis argued with careful reasoning that mitochondria are the remnants of ancient parasitic or predatory free-living bacteria belonging to the primitive order Rhodospirillaceae, part of a group known as alpha-proteobacteria. This group includes weird iron-precipitating microbes which form tiny chains of the iron mineral magnetite, and which have a similar appearance to nanoparticles sitting on the Martian meteorite ALH84001 found in the Allen Hills of Antarctica in 1984.[16] Like chloroplasts, mitochondria are fully adapted to life inside cells, where they too replicate autonomously to provide daughter cells with the battery packs they need to generate energy for fuelling their metabolism.

Equipped with its newly acquired chloroplast, the distant algal ancestor of green plants swam off into the strange oceans to change the world by transforming the seas and continents of the planet.[17] That transformation gathered pace around a billion years ago, when it gave rise to the two great lineages of plants: the Chlorophyta and Streptophyta (**Plate 1**). Each division is a large taxonomic grouping called a phylum and each followed its own spectacularly different evolutionary trajectory.[18] Chlorophytes are the green algae that diversified as plankton in the oceans and radiated into coastal and freshwater environments. They transformed the planet not by greening the land but by colouring the oceans; they are only distantly related to land plants. Chlorophytes are important, but streptophytes are the special group for our story. This group comprises the charophyte green algae, the first algae to conquer freshwater habitats, and land plants. Today's charophyte algae are not a particularly species-rich group, numbering a few thousand species at most. Their current diminished diversity reflects a progressive thinning-out by cataclysmic upheavals throughout Earth's history, especially during the mass extinctions at the end of the Permian and Cretaceous periods.[19] Modern charophyte species make up for this fact by encompassing a marvellous array of unicellular, filamentous, and complex multicellular forms that live in shallow or transient freshwater pools and streams, with a few able to survive in terrestrial habitats; damp soils are a favourite. It could even be that the charophycean green algae ancestor possessed a physiology allowing it to cope with the harsh terrestrial environment and had already been living on land for some time before the emergence of land plants.[20]

The close evolutionary relationship between charophyte green algae and land plants reveals something of singular importance: freshwater algae made the move from water to land and in doing so laid the foundations for modern plant diversity. It follows that freshwater is the ancestral habitat of the common ancestor of land plants. Plant life did not, as some popular accounts suggest, establish itself on land by 'storming the beaches'. Instead, this great event in biology began less dramatically, by the stealthy creeping of freshwater algal progenitors of land plants outwards through the soggy sediments of streams, rivers, lakes, and ponds. At first glance, quieter coastal environments might appear obvious places to make land. But success here would have demanded highly specialized floras, such as mangroves or saltmarsh plants.[21] Freshwater environments are altogether kinder to life, as we shall discover. Indeed, it is entirely possible that all fundamental diversification of plant life originated in freshwater, because red algae and the glaucocystophytes may also have evolved there, despite their current importance in marine environments.[22]

Botanists are, surprisingly, completely at a loss to explain what caused the deep evolutionary split within the green lineage creating the chlorophytes and the charophytes. Few have been brave enough to venture explanations for the schism but one hypothesis suggests the onset of 'Snowball Earth' conditions as a possible trigger.[23] Burkhard Becker of the University of Cologne is the leading exponent of this theory, and he points out that the two great lineages of green algae appeared during a series of major glaciations. Not for nothing is this interval of geological time called the Cryogenian. Geologists continue to debate whether the Earth could have frozen over completely to entomb the whole planet in ice each time, with life surviving under or above the great ice sheets, or whether habitable oceans existed in equatorial regions. Becker argues that the intense glaciations of the Cryogenian acted as environmental filters, selecting the best-adapted green algae from an ancestral pool of photosynthesizers living in different habitats at the time.[24] During the final glaciation of the Cryogenian, 580 million years ago, the major ice sheets once again expanded across the world, slowing the precipitation cycle. Freshwater pools and lakes dried up, forcing the evolutionary hand of algae: adapt to the challenges of a terrestrial existence and survive, or face extinction.

But why did charophyte and not chlorophyte algae give rise to land plants after the two lineages had diverged from one another a billion years ago? Both certainly had ample time, hundreds of millions of years, for expansion into new terrestrial environments. Part of the explanation could come down to simple matters of priority and ecological elbow room.[25] Charophyte algae colonized freshwater first and made land first. By the time chlorophyte algae had made their move into freshwater, any descendants evolving terrestrial tendencies would have faced fierce competition from plants already successfully adapted to the challenges of life on land.

Large red and brown seaweeds (which are large chlorophyte algae) colonized the shoreline and evolved complex life cycles, but they never really 'invaded' the land. Large fronds attached to rocky shores might seem like obvious progenitors of our land floras. Seventy years ago, this notion proved irresistible for the Oxford botanist Arthur H. Church (1865–1937). He proposed that algae from the intertidal zone were the starting point for the frontal 'assault' on land, but we now know they were no such thing. So taken was he with the whole idea that he postulated a hypothetical, and long extinct, green algal group that launched what he called a 'sub-aerial transmigration'.[26] Now regarded as little more than a botanical curiosity, Church's flight of whimsy reminds us of an important lesson: a proper understanding of evolutionary events demands a robust tree of evolutionary relationships.

Another reason for chlorophytes not making land is that there seems to be something intrinsically difficult about switching from the salty oceans to terrestrial habitats. Being pre-adapted to freshwater somehow predisposes you towards successful colonization of dry land.[27] It is perhaps easy to imagine the gradual movement of plants from freshwater to land by way of moist habitats that were subject to drying up from time to time and refreshed when the rains came.[28] Moving out of the salty oceans into freshwater, on the other hand, appears simply too dramatic and presents serious physiological challenges not easily overcome. Drastic metabolic adjustments are required to protect against the osmotic onslaught that ensues. If you take a freshwater species and submerge it in saltwater, life-sustaining moisture is drawn out through permeable membranes, causing death by dehydration. A saltwater species submerged in freshwater will succumb to the same osmotic action in reverse. Interesting evolutionary parallels can be drawn here between plants and animals that reinforce the idea. Insects originated in freshwater rather than marine environments, and amphibians and insects are still completely absent from the oceans.[29] Likewise, very few flowering plants moved into the sea after evolving from freshwater algae. Seagrasses are one of the few exceptions, and in so doing they shed parts of the genome for surmounting physiological obstacles to living on land.[30] A highly unusual group of flowering plants, seagrasses carpet the floors of shallow seas with strap-like leaves to form coastal meadows, with an astoundingly high level of productivity, comparable to that of fertilized maize fields or sugarcane plantations.

We are left with the charophyte algae as the closest living relatives of land plants, a fact regarded as 'proven beyond reasonable doubt'. The question of which charophyte lineage is the closest immediate ancestor of the earliest land plants is being hotly pursued. The answer would allow us to uncover the key features enabling such algae to make the important transition to a terrestrial lifestyle. In the 1980s, many botanists were convinced that a distinctive group of pondweeds called stoneworts (the Charales) were the closest living algal relatives of land plants,[31] with the Coleochaetales[32] ranking a close second. Rank outsiders, at that time, were a third group called the Zygnematales[33] (**Plate 2**). None of them is exactly a household name, and none trip off the tongue, so a few words of explanation are in order.

Forced to pick which of the above three groups of algae best resembles a proto-land plant, you would probably take a punt on the stoneworts.[34] This intriguing group, consisting of several hundred species worldwide, has cell walls that are heavily calcified and a tendency to become encrusted in calcium carbonate (limestone)—hence the name.[35] Their characteristic calcification has resulted in some closely related 400-million-year-old fossil specimens, beautifully preserved, with similar distinctive male and female reproductive structures.[36] Modern forms can look suspiciously more like vascular plants than green algae, with a central stem formed from giant cells linked end-to-end for tens of centimetres, and whorled 'branches' budding off at nodes along the stem. Dwelling in ponds, and more exotically in deep, clear volcanic crater lakes, these relicts are among the largest and most complex of all freshwater green algae and are often the first to colonize ponds that dry up in the summer. When the family tree of the charophytes is drawn based on inherited characteristics and their DNA, we find a satisfying stepwise evolutionary progression in complexity, moving from simple single-celled forms to more structurally complex multicellular forms appearing immediately prior to the origin of land plants.[37]

Elegant and apparently satisfying though this story may be, new large-scale DNA sequencing projects are radically revising our thinking about the origins of land plants, and suggest this earlier interpretation is quite wrong. Land plants instead likely arose from what are now far simpler algae,[38] with the finger of taxonomic suspicion now pointing to those within two ancient groups: Coleochaetales and the Zygnematales. Composed of microscopic branching filaments, sometimes equipped with flagella for swimming, Coleochaetales algae could not look more different from the complex Charales. Superficially, the Coleochaetales have few physical features in common with land plants, but they share hidden biochemical and structural features with them,[39] and a few modern forms can stubbornly survive and reproduce on mineral grains after a week or so out of water.[40] Fossils bearing a strong resemblance to some members of the living genus *Coleochaete* turn up in 400-million-year-old early Devonian strata, although these are substantially larger than any known extant representatives.[41] The last of the ancient groups of charophytes, and also the largest and most diverse of the living groups of green algae, the Zygnematales, look even less like land plants than the Coleochaetales, but are the current favourites to be the closest algal ancestors of plants. Forming single cells, filaments, chains of cells, and colonies, they have been observed under the microscope for their scientific interest and intrinsic beauty, and yet overlooked as possible progenitors of land plants, for over a century.

If the Zygnematales or Coleochaetales are most closely allied with land plants, we might wonder how the simple structure of their modern forms can be reconciled with complex terrestrial plants. It appears to run counter to our expectation that evolution produces complex life forms from simpler ancestors. The answer is that it is quite possible, probable in fact, that over time there have been reductions in the complexity of modern forms compared to their more complex ancestors. Since they last diverged from a common ancestor, each algal group has followed its own evolutionary trajectory for hundreds of millions of years, and in the case of the Zygnematales this may have led to a reduction in complexity of modern forms. So we should be mindful that the appearance of the modern forms we see today can be misleading.

Nevertheless, probably by virtue of being in the right place at the right time, charophyte algae gave rise to the first plants that ventured onto land to make a living beneath the sky. As those elementary grades of land plants slowly took hold, creeping over muddy sediments, they flecked the desolate, windswept valley floors with new shades of green. The closest surviving relatives of those early land-seeking pioneers are simple plants called the bryophytes, whose life history and biology open a window on their long-extinct ancestors.[42] The bryophytes comprise the liverworts, mosses, and hornworts. These organisms are typically small plants, reaching a few centimetres in height, and grow under damp conditions. Some are tolerant of desiccation and able to endure extreme heat, like our Mojave Desert mosses. The evolutionary relationships among the bryophytes, and between them and vascular land plants, are controversial. Almost all possible permutations have been proposed at one time or another, a situation reflecting what is widely regarded as one of the most recalcitrant and frustrating problems in land plant evolutionary biology.[43]

Regardless of the confusion surrounding the kinship of bryophytes, there is a close resemblance between the fossilized spores of the earliest land plants and those of liverworts. Charles Wellman of the University of Sheffield made the discovery after his detailed investigations of ~460-million-year-old Ordovician rocks from Oman.[44] His fossils included spores and remarkable fragments of spore-producing plants in the form of specialized reproductive structures called sporangia, still loaded with spores. Slicing through the fossil spores, he discovered that the wall is constructed in distinctive layers, sandwiched together in a way similar to that found in modern liverwort spores. The oldest fossilized bryophytes are also classified as liverwort-like, and are older than those thought to represent mosses and hornworts by hundreds of millions of years.[45] We have thick mats of fossilized liverworts excavated from 380-million-year-old sediments by quarrying in New York State[46] and a rare liverwort specimen in rocks from South-west China dated as 411–407 million years old.[47] These discoveries provide direct fossil evidence for advanced groups of liverworts on land over 400 million years ago.

And there, frustratingly, the trail of fossil evidence runs cold, but a clue to what the mysterious earlier forms might have looked like comes from the latest taxonomic hierarchy of liverworts, identifying the extreme antiquity of the Haplomitriopsida. Comprising just three genera, *Treubia, Apotreubia,* and *Haplomitrium,* this class of primitive liverworts may have an ancestry reaching back 450 million years.[48] Strange-looking plants, they are quite unlike the 400-million-year-old fossilized liverworts discovered in the USA and China. Those of the genus *Treubia* have prostrate green shoots adorned with small leaf-like structures folded upwards like miniature wings, giving the plants a ruffled appearance. Those in the other significant genus, *Haplomitrium,* are different. They grow with subterranean rhizomes and erect stems adorned with rounded leaf-like structures.[49] Could these plants, belonging to the least celebrated and oldest living liverwort lineage, be furnishing us with a tantalizing glimpse of the earliest terrestrial plant life?

Fossil spores extracted from ancient rocks hint at the nature of land floras back in the Ordovician (485–444 million years ago), when simple proto-bryophyte plants began to give Earth's continents a patchy green photosynthetic veneer.[50] But does the record of fossilized spores accurately date when plant life began to make tentative moves on to land? As we have seen, this would have been a subtle event, involving very simple photosynthesizing life forms that had evolved from algae. Little, if anything, of this change may have survived in the fossil record. Those first land plants still had to become sufficiently numerous and widespread to be captured by accidents of fate and preserved in stone. Fossils therefore tend to post-date the actual time of land-plant origination by some unknown duration, usually providing what is regarded as the youngest age estimate for when land plants appeared. The 'unknown duration' constitutes a 'known unknown' and is contingent on measures like the ecology of the organisms, their preservation potential, and the nature of the rock record.

Philip Donoghue at Bristol University elegantly addresses these sorts of seemingly impossible questions by exploiting the clock-like properties of DNA molecules.[51] Nobel Prize-winning chemist Linus Pauling (1901–1994) and his colleague the Austrian-born French biologist Émile Zuckerkandl (1922–2013) originally proposed the concept while working on haemoglobin molecules back in the 1960s. It is based on observations showing that genetic mutations, although random, occur at a *relatively* constant rate, which means that the number of differences between any two gene sequences increases at a predictable rate over time. At least in theory. In other words, instead of measuring time in the usual units of seconds, hours, days, and so on, molecular clocks measure time as the number of mutations in a particular gene sequence. As with regular timepieces, molecular clocks have to be calibrated, and this is usually done with a dated fossil specimen for a specific lineage. Once the rate of mutation is determined, calculating the time of divergence of that species is relatively straightforward. If the rate is 5 mutations every million years, and you count 25 mutations in your DNA sequence, then your sequences diverged 5 million years ago.

Of course, there is more to it than this, but the basic idea holds true, and Donoghue is a whizz at crunching complex numerical algorithms with computer software. Molecular clocks, calibrated by fossils, give a timescale by which the plant tree of life has unfolded that may help in overcoming the unreliability of fossils for dating the origin of land plants. If Donoghue's team have a sensible grasp on Pauling and Zuckerkandl's molecular chronometers, the sensational outcome of their latest work is to propose nothing less than redrawing the timeline of green evolution.[52] Their suggestion is that plant life emerged onto land sometime in the middle Cambrian–early Ordovician world (515–470 million years ago). We have no firm fossil evidence for land-plant existence towards the older end of this range, as his team readily acknowledges, but should this be an obstacle to accepting the DNA evidence? Wellman gives the idea short shrift, arguing that his spores are indestructible and once land plants began reproducing by wind-blown spores they would soon be dispersed and turn up in marine and terrestrial rocks. Could a dusting of such spores around the planet have been hidden from rocks cleaved by the blows of a geologist's hammer or their vats of rock-dissolving hydrofluoric acid for so long? Wellman, and his palaeobotanical colleagues who collaborate with Donoghue, doubt it, but the fact is few unequivocally terrestrial rock formations that could document the conquest of the land by plants much older than 450 million years are still in existence.[53] The upshot is that Donoghue's controversial early date for the origin of land plants is hard to test[54] but it has prompted palaeobotanists to turn their attention towards the more complete geological record of marine rocks that might capture crucial evidence of the tough powdery spores. If we lean towards the younger end of the range, it fits together with dates for the arrival of animals that fed on plant remains, such as arthropods and their relatives,[55] but that date is also obtained using molecular clocks with algorithms suffering similar issues of uncertainty. Understandably, the prevailing opinion is one of scepticism. Yet ancient Scottish deposits have yielded fossil discoveries pointing to the existence of freshwater algae in pools a billion years ago.[56] Were ancient freshwater algae plus a few hundred million years sufficient to boot up simple forms of terrestrial plant life?[57]

Regardless of the exact timing of the origin of our land floras, once plant life established itself on land, the stage was set for evolutionary escalation; there was no turning back. The green fuse of land-plant evolution was lit, and was not about to be extinguished anytime soon. The pull of a terrestrial lifestyle proved irresistible. Limitless supplies of solar energy raining down from the sky and a luxurious carbon dioxide-rich atmosphere fuelled photosynthesis.[58] Despite the challenges of coping with intermittent supplies of water and nutrients from thin primitive soils, the 'pull of the photon' made living on land irresistible. Endless ecological possibilities offered by the open terrestrial landscape were soon exploited as successive waves of vascular plants followed the early photosynthetic pioneers in populating the continents.[59] Vascular plants appeared with their hallmark stiffened stems, with internal tubes conducting water and dissolved mineral nutrients taken up from the soil. Some vascular lineages rose to great ecological success, shining brightly in the terrestrial world for tens of millions of years before fading and being replaced by other more successful groups. The story is chronicled by DNA and fossils, precious stony way-markers of the directionality of plant life's emboldened exploration of the terrestrial realm.

The living descendants of this botanical drama are still with us in the shape of clubmosses and their relatives (lycophytes) and ferns (pteridophytes). Lycophytes have the distinction of being the oldest living lineage of vascular plants. Their fossil history is thought by some to reach back to a fossil plant called *Baragwanathia*, controversially dated to 420 million-year-old Silurian strata in Australia. Fossilized shoots of *Baragwanathia* bear close comparison with those of its modern lycophyte relatives. Lacking true leaves or roots, and reproducing with spores rather than seeds, the lycophytes represent an interim step in plant evolution. Stems were cloaked in small scale-like leaves adorned with microscopic stomatal pores, and underground parts of the plants were anchored into the fine substrate by simple rootlets.[60] Supported by woody or lignified shoots, vascular plants began to grow upwards in the struggle for light. Long-extinct relatives of lycophytes grew to be giants, reaching magnificent heights of over thirty metres. They towered over the primordial Carboniferous swamp forests 300 million years ago.

Other groups of fern-like plants soon appeared, joining the burgeoning diversity of bryophytes and lycophytes in greening the landscapes of the Devonian world. With upgraded leaves for harvesting solar energy, and stems plumbed into the soil by roots penetrating between soil particles to absorb films of water and dissolved nutrients, these opportunistic colonizers prospered. Ferns diversified rapidly into prostrate crawlers, climbers, scramblers, and sprouters. Some occupied shady quarters of the forest floor and others colonized flood plains, regions decimated by wildfire, and the flanks of volcanoes. By far the greatest diversity sprang up in the tropics.[61] Ferns reproduced by releasing spores from orange-brown packets located on the underside of fronds. Some of the stranger examples in our modern floras are the whisk ferns like *Psilotum* and the horsetails (*Equisetum*). The stems of horsetails are encircled by rings of needle-like leaves that look quite unlike true ferns, and have spore-bearing cones on the tips of the shoots. Appearances can be deceptive, however. For many years, classifications based on morphology (the look of the thing) saw them separated from true ferns, only for the molecular characters of their DNA to reveal that whisk ferns and horsetails are actually true ferns, no matter how strange they look.[62] Like their lycophyte cousins, the direct ancestors of horsetails reached lofty heights in the Carboniferous, at a time when giant dragonflies took to the wing.[63]

New shades of green and a rich diversity of leafy textures followed, as seedless vascular plants began proliferating and filling out the landscape with angular new geometries. Really, though, it was the origin of seed plants that ratcheted up the unstoppable business of greening the planet, giving it significant extra energy and momentum. The first seed plants arose late in the Devonian, at least 365 million years ago, and a wide array of seed-bearing plant forms followed. All were probably descended from an extinct seedless group of plants called the progymnosperms that foreshadowed the forests that were still to come. Some members, like the tree *Archaeopteris*, attained heights of 17 metres or more, and occasionally formed extensive stands of trees. These early seed plants ultimately led to the appearance of gymnosperm trees that went on to dominate forests throughout the age of the dinosaurs, the Mesozoic Era.

Just four living groups of gymnosperms survive in our modern floras—conifers, distinctive gnetophytes, cycads, and the maidenhair tree (*Ginkgo*). Gymnosperms (from the Greek *gymnospermos*, meaning 'naked seeds') develop seeds either on the surface of scale-like appendages of cones or at the end of short stalks in the case of *Ginkgo*, with its distinctive fan-shaped leaves. Conifers are the largest group, with over 600 species, including the pines, spruces, larches, and firs that dominate vast tracts of boreal forest in the subarctic regions of North America and Eurasia. Gnetophytes are weird plants with features similar to conifers and flowering plants. None is stranger than *Welwitschia*, named after the Austrian botanist Friedrich Welwitsch (1806–1872) who discovered it in 1859. Confined to the Namib Desert in southern Africa, *Welwitschia* can live for decades or even centuries. Described as the 'plant from Mars', it is not hard to see why. The plants live a curious existence, persisting with two sprawling strap-like photosynthetic leaves that inch across the desert floor often reaching ungainly lengths of several metres. Cycads are the second most diverse and widespread lineage of gymnosperms, with about 300 species. They are distinctive for the dense whorl of leathery leaves sprouting near the top of a thick trunk. At first glance, cycads might be mistaken for small palm trees or even ferns, until you notice the striking heavy seed-bearing cones topping these sturdy trees, which once flourished amid the dinosaurs during the Jurassic. Today, they comprise relict components of our floras, representing the last remnants of their formidable past.[64]

Understandably, Michael Crichton could not resist the charms of cycads in the film, *Jurassic Park*. 'Those trees to your left and right are cycads, prehistoric predecessors of palm trees. Cycads were a favourite food of the dinosaurs', intoned the narrator as the occupants of the land cruisers drove through the park. We can admire the botanical correctness of placing cycads amongst the dinosaurs but should note that scientists from the Natural History Museum, London, have since revised the story of cycad-eating dinosaurs.[65] Once, the thick, tough stems and leaves and the extreme toxicity of the foliage were thought to have evolved as protection against dinosaurs. Sadly, this romantic notion has fallen to modern science. The toxicity of cycad foliage evolved sometime in the Permian, tens of millions of years before the dinosaurs appeared, probably as a by-product of primitive plant biochemical pathways. The idea that brightly coloured cycad seeds appeared as a means of attracting dinosaurs as agents of dispersal has also fallen to the same team from the Natural History Museum—fossilized dino-dung shows no trace of cycad seeds. The job of cycad seed dispersal back then fell to other exotic Cretaceous fauna, and to oceanic currents in the case of those lineages clinging to coastal regions.

Ascending towards the upper branches of the plant tree of life, we leave the gymnosperms behind to encounter the angiosperms, or flowering plants, which burst onto the evolutionary stage, transforming the surface of the Earth with colour. The flower power revolution took off early in the Cretaceous, around 125 million years ago. Molecular clocks suggest that flowering plants originated far earlier and indeed there have been scattered reports for their putative remains excavated from Triassic and Jurassic rocks. None of these claims, however, has yet stood up to critical scrutiny.[66] Today, there are more than 223 000 species of flowering plants, dwarfing the diversity of all other plant lineages; 75–85% of all known plant species on Earth are flowering plants (**Figure 2 and Plate 3**). Molecular sequencing, combined with the rich fossil record of flowers, seeds, and leaves, has proved successful in resolving much of the gloriously detailed family history of the two main evolutionary lines—the 'monocots' and the 'eudicots'.[67] The monocot group, including lilies, orchids, sedges, and the grasses that gave us cereals, has only one seed leaf. Seedlings of the eudicots typically develop a pair of tiny green leaves quite distinct in shape from those of the mature leaves that develop subsequently. Most of the familiar trees, shrubs, and herbs in our woodlands and hedgerows are classified as eudicots.[68]

Writing for the *National Geographic*, the American author John Updike (1932–2009) paints the natural history of the scene on land for us sometime late in the Cretaceous, when many of these groups of plants had evolved, like this:

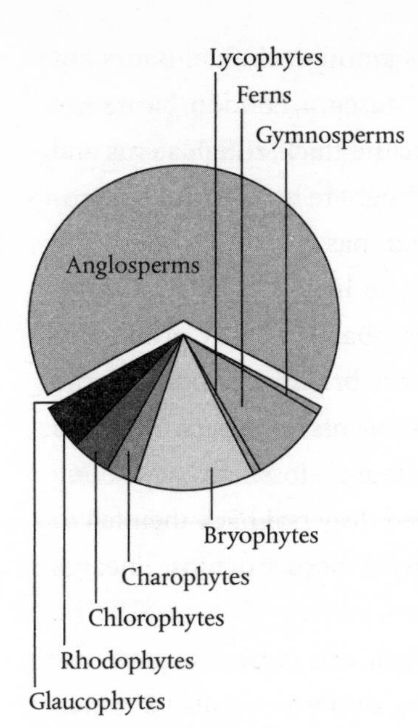


**Figure** 2 The relative species richness of the major clades of plants.

> By the late Cretaceous the continents had something like their present shapes, though all were reduced in size by the higher seas, and India was still an island heading for a Himalaya-producing crash with Asia. The world was becoming the one we know: the Andes and the Rockies were rising; flowering plants had appeared, and with them, bees. The Mesozoic climate, generally, was warmer than today's, and wetter, generating lush growths of ferns and cycads and forests of evergreens, ginkgoes, and tree ferns close to the Poles; plant-eating dinosaurs grew huge, and carnivorous predators kept pace. It was a planetary summertime, and the living was easy.[69]

The living was easy, of course, because green plants had won the land, no question, and become fabulously successful at the business of reproducing on dry land. The stars of those early green waves of colonization, the proto-bryophytes, bryophytes, lycophytes, and ferns, all reproduced with spores and not seeds, and remained dependent on water for fertilization. The same holds true for the strugglers and the also-rans in land-plant evolutionary history, represented by extinct vascular plant groups, including obscure groups such as the rhyniophytes, trimerophytes, and progymnosperms. All had acquired the ability to reproduce on land thanks to a dramatic shift in their life history that provided an enormous advance over the watery sexual mechanisms of their algal ancestors.

Take the weird sex life of the filamentous photosynthesizer algae of the order Zygnematales, a clue to which is in their common name—the conjugating green algae. Unlike most other green algae, they lack flagella and reproduce sexually through variations in the process of conjugation. Cells of opposite gender line up in parallel filaments and tubes form between corresponding cells. Strangely, the non-flagellated male cells then become amoeboid and crawl across to meet the egg cell. Here they fuse and form a zygote, which later undergoes cell division to produce offspring, with only the female passing chloroplasts on to her offspring. The business of reproduction is no less strange in the Coleochaetales. Bathed in a supportive and protective watery environment, adult plants develop specialized structures that produce a large egg and smaller free-swimming male sperm. Sperm fertilize the egg to form a zygote, which is retained on the parent plant and undergoes cell division to produce spores. Cells surrounding the zygote divide to produce a layer of sterile tissue that envelopes it, and possibly provides it with nourishment during spore production. Finally, the zygote ruptures and releases its cargo of swimming spores. We can see from this outline that *Coleochaete* is obviously totally dependent on water for sexual reproduction. Yet, at the same time, it has the beginnings of something resembling care of a developing embryo, one of the hallmarks of land plants.

From this rather mixed bag of algal reproductive biology, land plants evolved a critical modification to their life history that would change the world. In some distant charophyte ancestor, after fertilization and the formation of a zygote, something different happened. Instead of directly producing spores, the cellular machinery was reprogrammed to produce a tiny erect multicellular plant, the 'sporophyte', for manufacturing spores. This small but momentous event gave plant life a new distinction between the 'sporophyte' and the parent plant that produced the eggs and sperm (gametes), called the 'gametophyte'. The sporophyte grew, at least initially, attached to its amphibious parent, anchored into 'soggy sediments'. The first such structures developing on tiny ancient green plants would hardly have qualified as *bona fide* 'land plants'. While details of the life cycles of the proto-bryophytes that first took hold on the land are sketchy, some simple variant of this basic advance probably led to their persisting out of water for several generations.[70] Merciless natural selection ensured only the fittest surviving descendants would live on to reproduce and turn the continents green. Land ahoy.

This development represented a crucial and profound step forward, but at first the advantage of nurturing a multicellular sporophyte would have been subtle. The immediate benefit of sporophyte generation was to facilitate the production of millions of spores following a single fertilization event. And by growing above the ground surface, plants could release spores into more turbulent air currents found there, ensuring they travelled further and spread out to colonize new areas. In the longer run, the sporophyte phase became free to grow larger, more complex, and develop specialized tissues for survival and spore production and dispersal on land. The fossil record tells us it was not long, probably a few tens of millions of years, before land plants evolved large size differences between the gametophyte and sporophyte generations. By around 410 million years ago, sporophytes had advanced to become an autonomous generation, free of their smaller fleshy gametophyte parent that attached itself to muddy substrates by simple rootlets called rhizoids.[71] Ever since, across the grand sweep of plant evolutionary history, the general trend has been one of an increasing dominance of the sporophyte generation at the expense of the gametophyte generation. The natural histories of bryophytes, ferns, and seed plants illustrate the trajectory that ultimately saw the reduction of the gametophyte generation in flowering plants to a less-than-conspicuous handful of cells.

All three groups of bryophytes (mosses, liverworts, and hornworts) have life cycles reflecting those early stages of terrestrial events, with a dominant gametophyte generation and diminutive sporophyte generation. Take the life cycle of the prostrate liverworts. Sexual reproduction begins in the gametophyte generation when the flattened thallus of the male plant sprouts miniature umbrella-like structures that produce sperm, and the females sprout umbrella-like structures that make the female reproductive organs. For successful fertilization, and to avoid inbreeding, some liverworts have evolved remarkable tricks. Microscopic sperm released by the male plants are large and have flagella, making them strong swimmers. Often they are released simultaneously with lipid droplets, which aid their dispersal by carrying them on a lipid 'skin', perhaps only a single molecule thick, that spreads out over ten metres. The films of water carrying the sperm ascend the umbrella handles of the female plants by capillary action, aided by narrow grooves and little pegs to ensure water rises and delivers the sperm to the eggs. When sperm fertilize the female egg cell, the embryo plant or sporophyte generation develops. Protected in a sac suspended from beneath the stalk of the umbrella, the miniature multicellular plant draws its nutrients from the female parent plant.[72] Wrapped in a tough coating, the spores it makes gain protection from the elements and are released when gentle currents of air or rain drops buffet the tiny plant structure. When conditions are right, the spores germinate to produce a new thallus. On reaching maturity, up goes its umbrella to complete the life cycle.

Details differ between bryophytes, and the life cycle described above is quite an intricate, advanced example. Mosses provide a simple example (**Figure 3**). In suitable conditions, spores germinate to give rise to a filamentous thread (protonema). Some of these give rise to the rhizoids, stems, and leaves that make up the adult

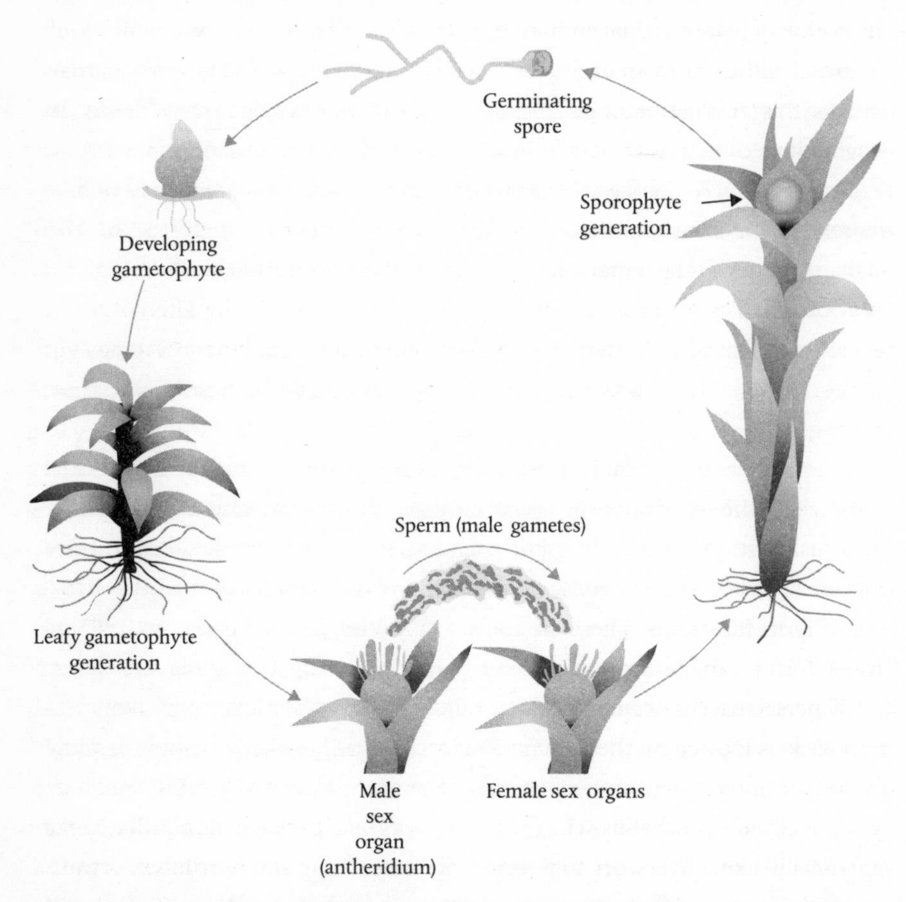


**Figure 3** Lifecycle of a moss (*Physcomitrella*) with an alteration of generations between the gametophyte and the sporophyte phases.

gametophyte (also called the leafy gametophore) which carries the male (antheridium) and female (archegonium) sexual organs (together called the gametangium). The antheridium produces flagellated sperm that actively swim in a film of water to the archegonium for fertilization. The resulting fertilized egg cell (zygote) gives rise to the embryo, the sporophyte represented here by the spore capsule (sporangium).

The critical point is that it illustrates how a single species of plant can have two generations that are completely different to each other. The insertion of a multicellular sporophyte generation was a fundamental developmental innovation of land plants and allowed nurturing of the developing embryo, a defining feature of the biology of all land plants (which is why botanists call them embryophytes). The notion of plants as dual entities, having two distinct generations, which look completely different from each other, takes a bit of getting used to. No animals undergo this 'alternation of generations'. Not that this has been a barrier to the imaginations of fiction writers—think of Robert Louis Stevenson's 1886 classic *The Strange Case of Dr. Jekyll and Mr. Hyde*, or Marvel Comics' The Incredible Hulk—where very different manifestations are imagined for the same creature. You might imagine that the remarkable transformative metamorphosis of caterpillars into butterflies is an example but it is not a true case of the alternation of generations. The adult butterfly appears to act like a sporophyte by laying eggs but the caterpillar is not acting in the role of a gametophyte, with male and female sex organs.

By the time early vascular land lineages, such as lycophytes and ferns, appeared on the scene, the sporophyte generation had undergone significant evolutionary advances. It was no longer diminutive, and no longer dependent on its parents. Instead, it had become a large, dominant, and free-living plant, as the natural history of ferns illustrates. There are more than 10 000 fern species, with highly diverse forms, ranging from minute understory herbs to spectacular tree ferns. The dispersal of ferns occurs largely through the release of millions of tiny spores from packets located on the underside of arching fronds. Those spores, landing in a suitably moist environment, germinate into a small, filmy, heart-shaped sexual plant called a prothallus. The prothallus is a small package of cells that looks superficially like a liverwort, and which produces male and female sex organs. This is the gametophyte generation. Cells on its underside release male sperm into a film of water that allows them to wriggle towards the egg cell and fertilize it. After fertilization, an embryonic fern develops by producing a little leaf above the gametophyte, and a little root below. Initially the young plant is supported nutritionally by the gametophyte but it quickly becomes independent. After the development of a few more leaves and roots, it grows into an adult plant, complete with graceful fronds for spore production. In ferns and clubmosses, this adult plant is, of course, the sporophyte generation. It has become a large, free-living entity, able to shed millions of spores into passing air currents. But that freedom comes at a price. The prothallus required for sexual reproduction remains small and confined to damp soil, where it is vulnerable to being eaten, drying out, or starvation if neighbouring plants interpose and cut out life-giving sunlight.

Seed plants—the gymnosperms and flowering plants—solved this and the other problems facing the ferns and their kin by moving the locus of sexual generation up into the canopies of the forest. This opened up a new range of possibilities for ensuring the sex cells could meet each other without the need for water-based transportation. Take the conifers, for example: all have the same basic life cycle, involving the production of distinctive male and female cones. Male cones manufacture masses of wind-blown pollen grains and the female cones hold the eggs. When a pollen grain lands on a female egg-bearing cone a remarkable feat of endurance ensues, the individual grain slowly growing a pollen tube down towards the egg, sometimes taking a whole year to reach its final destination. Once it reaches the egg, the tube releases its sperm cell, and fertilization occurs without the need for water. The fertilized egg remains on the cone for a further year. During that time, rich food supplies are laid down around it, and it is further wrapped in a protective coat to form a seed. More than two years after fertilization, when the cone dries, the segments slowly creak open, releasing the seeds, which await conditions suitable for the germination of a next generation. Conifers had at last eliminated the need for water to transport sperm for fertilization.

These points were not lost on the American anthropologist and gifted natural science writer, Loren Eiseley (1907–1977), who wrote with an almost poetic sensibility on plant evolution in his collection of essays entitled *The Immense Journey*.[73] Selling over a million copies at the time of publication, the book become a runaway bestseller to an adoring public on both sides of the Atlantic. In a single paragraph, he provided a neat summary of the immense journey of the title:

> Slowly, toward the dawn of the Age of Reptiles, something over two hundred and fifty million years ago, the little naked sperm cells wriggling their way through dew and raindrops had given way to a kind of pollen carried by wind. Our present-day pine forests represent plants of a pollen disseminating variety. Once fertilization no long required exterior water, the march over drier regions could be extended. Instead of spores, simple primitive seeds carrying some nourishment for the young plant had developed.

Successful though the conifers undeniably are, they rely on wind-blown pollen for sexual reproduction, and this is a haphazard and biologically expensive business at the best of times: the majority of the pollen that trees release to the winds is wasted, failing to land on female egg-bearing cones. Prodigious quantities of pollen must be made, each cone manufacturing millions of grains; tap a male pine cone in spring and great clouds of golden dusty pollen diffuse into the air with nothing to fertilize. The solution to this problem, evolved by the angiosperms, was both elegant in its effectiveness and beautiful in appearance. That innovation was, of course, the flower.[74]

Rather than producing masses of pollen and releasing it into the wind to take its chances, flowers provide receptacles containing male pollen-producing organs and female egg-containing organs for receiving the pollen of other flowers. Unlike modern flowers, those of the earliest flowering plants were probably small and inconspicuous, and pollinated by the wind and insects, though it is difficult to be sure about this.[75] Later lineages evolved flowers that shifted the balance of power towards plants. They attracted insects as animal couriers that would make precise simultaneous collections and deliveries of pollen. By producing scent and brightly coloured petals to advertise the rewards on offer, flora gained a certain mastery over fauna. Fertilization of a flower is independent of water. It leads to the development of an embryo encased in a seed supplied with its own energy reserves.[76] What could provide a better start in life?

Notice that, by this stage of plant evolution, the gametophyte generation in flowering plants has dwindled to virtually the smallest possible biological denominations: the male version is simply the pollen tube with sperm cells; the female version is a handful of cells containing the egg. They are so small that botanists invented a new term for them—microgametophytes. So an 'immense' evolutionary journey that started with gametophytes as the earliest dominant multicellular green life forms on land reached a point where this phase of the life cycle is subsumed into the sporophyte generation. The sporophyte generation now accounts for most of the amazing diversity of photosynthetic life forms on land.

With the evolution of seeds and flowers, reproduction changed forever, and the need for water to achieve fertilization finally became superfluous. As biological inventions, both flowers and seeds are fascinating. This is not the place to become side-tracked by the marvellous array of specialized devices flowering plants evolved to trick insects, birds, and mammals into pollinating them. Or, come to that, to prevent self-pollination. Instead, here is Loren Eiseley again, this time providing a charming vision of the rise of flowering plants:

> The ramifications of this biological invention were endless. Plants travelled as they had never travelled before. They got into strange environments heretofore never entered by the old spore plants or stiff pine cone seed plants. The well-fed, carefully cherished little embryos raised their heads everywhere. Many of the older plants with more primitive reproductive mechanisms began to fade away under this unequal contest.

The explosive diversification and global ecological takeover of flowering plants constitutes one of the most extraordinary and far-reaching bursts of plant diversification ever witnessed on Earth, and it happened within an astonishingly brief window of a few million years early in the Cretaceous.[77] In the tropics, forests of angiosperm trees sprang up whose dense, complex canopies filled with animals such as butterflies and beetles exploiting the opportunities on offer high above the forest floor. The corpulent bees in Updike's description of the Cretaceous diversified simultaneously with the evolution of specialized pollination modes by flowering plants.[78]

By around 40–30 million years ago, another group of plants began to stir: the remorselessly successful grasses.[79] Grasses and grassy habitats are now everywhere you look—think of the vast temperate grasslands such as the Great Plains of North America and the tropical savannas. Think also of the grasses that feed the world—rice, wheat, corn, barley, rye, and oats. Grasses owe their success, in part, to a remarkable feature of their biology: an ability to sprout new leafy shoots from underground corms and rhizomes. It ensures they can survive and regenerate following wildfire and makes them resistant to grazing. The spectacular rise of grasslands across the continents was driven by climatic deterioration throughout the Cenozoic (the past 65 million years) as the great ice sheets of the world formed in the polar regions, locking in water and increasing global aridity and the incidence of fire. Humans have increased fire and grazing over the past 10 000 years of the Holocene epoch, and accelerated grassland expansion by clearing land to make it available for growing crops and the grazing of domestic animals.

In the lower latitudes, grasses with a specialized photosynthetic pathway originated as their ancestors moved out of the shady tropical forests and into the bright, open, sunlit world of tropical woodlands and savannas.[80] These are the $C_4$ grasses, a name that derives from details of the photosynthetic biochemistry which turbocharged their productivity.[81] Indeed, the most productive plant species in the world is a tropical $C_4$ forage grass called *Echinochloa polystachya*.[82] Modern $C_4$ grasslands account for nearly a quarter of the terrestrial primary productivity on the planet. The oldest $C_4$ grass lineage is the Chloridoideae, which may have originated nearly 30 million years ago.[83] By far the majority of $C_4$ grass lineages originated more recently than this, smouldering on a long fuse before diversifying to dominate the ecological stage within a few million years during the Miocene, around 7–8 million years ago.[84] As fertile grassy savannas and plains around the world filled with $C_4$ grasses, they began forcing the evolutionary hand of herbivorous mammals; in came herds of grazers, out went the browsers and the nibblers.

Elsewhere, in the deserts of the globe, at around the same time as the grasses were conducting their grassy revolution, other plant forms like cacti and ice plants—the succulents—began seizing the initiative. Once upon a time, the history of desert vegetation was a closed book. The potential for fossilization of plant life in such inhospitable regions is almost non-existent; conditions are too dry to favour organic preservation in the rock record. Cacti, like the epiphytic ferns, have left little behind in the way of fossils to tell us of their evolutionary history. Once again, though, analyses of the DNA of living cacti have uncovered how old these succulents are, and when they began populating the deserts of the world. Before DNA evidence, we knew little about when they evolved and estimates of the time of their emergence ranged from 90 million years ago to 14 million years ago. DNA sequencing of cacti now reveals the group to be about 35 million years old. They underwent a dramatic burst of diversification synchronously throughout the deserts of the New World late in the Miocene.[85] Old World succulents, ice plants and their kin, diversified around the same time too. These radiations are tied to geological events, such as the uplift of mountain ranges, and the opening and closing of seaways, triggering the expansion and intensification of arid climates.

We can see now that a comparatively narrow window of Earth history witnessed the contemporaneous spread of specialized grasses and succulents across the global landscape, driven by a relentless exploitation of new ecological opportunities created by expanding arid climate regimes. By around 10 million years ago, Earth's spectacular transition to a grassy planet, with deserts populated by cacti, marked the final stages of the evolutionary assembly of modern-day floras. Land floras recognizable as those resembling their modern forms had finally arrived. What started half a billion years ago, when the descendants of freshwater green algae moved on to land, triggered an unstoppable succession of ferns, forests, and flowering plants that transformed the continents. Of course the cryptobiotic crusts where it all began are still with us, and doing nicely thank you, living in extreme environments. They remind us that one form of plant life is not simply replaced by another. The modern plant kingdom draping the continents in colour is an astounding evolutionary legacy stretching back to green algae, and further back still, to the momentous capture of a cyanobacterium nearly two billion years ago. That cellular hostage-taking event handed the distant ancestors of all plant life the ability to photosynthesize. Nothing like it has ever happened since, and plants never looked back.

In the end, the conquest of the land by green plants is really a story about an extraordinary developmental innovation that produced an extra step in their life cycle. This is the multicellular embryonic plan that we call the sporophyte generation. It provided the platform for all that followed, building bigger, more complex, and more productive plants, and ultimately emancipating them from the vagaries of water supplied by the weather gods for reproductive success. We would probably all concede that 'modern parents hover at about A-level standard on the prehistoric fauna' while 'children are all PhDs'.[86] Yet when it comes to appreciating prehistoric floras, their origins, and how our modern floras came to be, adults and children alike are firmly rooted at kindergarten level. The green tree of life constructed from fossils, DNA molecules, and a close reading of the biology of living plants, reveals the vanished ecology of the green world's evolutionary history. It helps redress the plant–dinosaur balance. Instead of simply seeing plant life around us, we begin to grasp a sense of the complex and extraordinary evolutionary trajectory that brought it to this point, and to appreciate the importance of these developments for our own evolution and survival, and that of all other animal life on the planet.

In Chapter Three, we discover how the genomics revolution of the past decade is revealing unprecedented insights into the major transitions in the evolutionary history of our land floras, as a riot of green diversity exploded across the continents.

# 3 GENOMES DECODED

'The genome revolution is only just beginning.'

J. Craig Venter, 2010, *Nature*, **464**, 676–7

In 2008, scientists announced they had obtained the partial genome sequence of long-extinct woolly mammoths using DNA extracted from the hairs of a mummified mammoth buried in Siberian permafrost for at least 20 000 years.[1] A year later, they trumped this remarkable achievement by reporting the genome sequence of Neanderthals, the closest relatives of modern humans, who once lived in Europe and western Asia before disappearing 30 000 years ago.[2] Inevitably, the world's press seized on the *Jurassic Park*-like possibility of resurrecting mammoths and Neanderthals by mining these DNA strands and filling the gaps with genetic material from the next best thing, elephants and humans. Plant biologists cannot yet make these claims for such iconic forms of past life on Earth, but smile at the trivial distances back in time the studies reach. Genome sequencing of photosynthetic life forms at the base of the green tree of life takes us back nearly two billion years to the origins of photosynthesis itself.

The genome of an organism is simply the entire DNA content packed inside its cells. Sequencing means determining the exact order of the four chemical building blocks ('letters' or, more technically, bases) from which the DNA molecule is constructed. The DNA molecule is a double-stranded affair, with specific bases on each strand pairing together opposite each other in a manner strictly determined by chemical rules of attraction and repulsion. These letters code for small molecules known as amino acids. Strung together into long-chain molecules to make proteins, amino acids are the building blocks of life. Each sequence of amino acids yields a different protein. The staggering truth is that variations in a genetic alphabet of just four letters generate the enormously rich diversity of life on Earth.[3]

It is worth pointing out early on that it would be a mistake to think of the genome as a series of genes linked together, immediate neighbours residing next door to each other. The reality is that genes are just a small fraction of the total DNA of a genome. Often genes are separated by vast tracts of repeated DNA sequence, whose function (if any) is not yet clear, and genes themselves can be split into multiple parts. In fact, genes, and those snippets of DNA that regulate gene expression, typically account for only a small fraction of total genome size. The rest of it is composed of stretches of repeated sequences that have accumulated over millions of years. These sequences, known as mobile DNA elements, can move around their host genome and multiply via 'copy and paste' or move via 'cut and paste' mechanisms. Over time, their activities generate huge variations in genome size without apparent correlation with organismal complexity (**Figure 4**).[4] This helps explain Steve Jones' remarks in his book *Darwin's Island* that 'a chicken has slightly less DNA than a Nobel laureate', and that the 'tiny plant

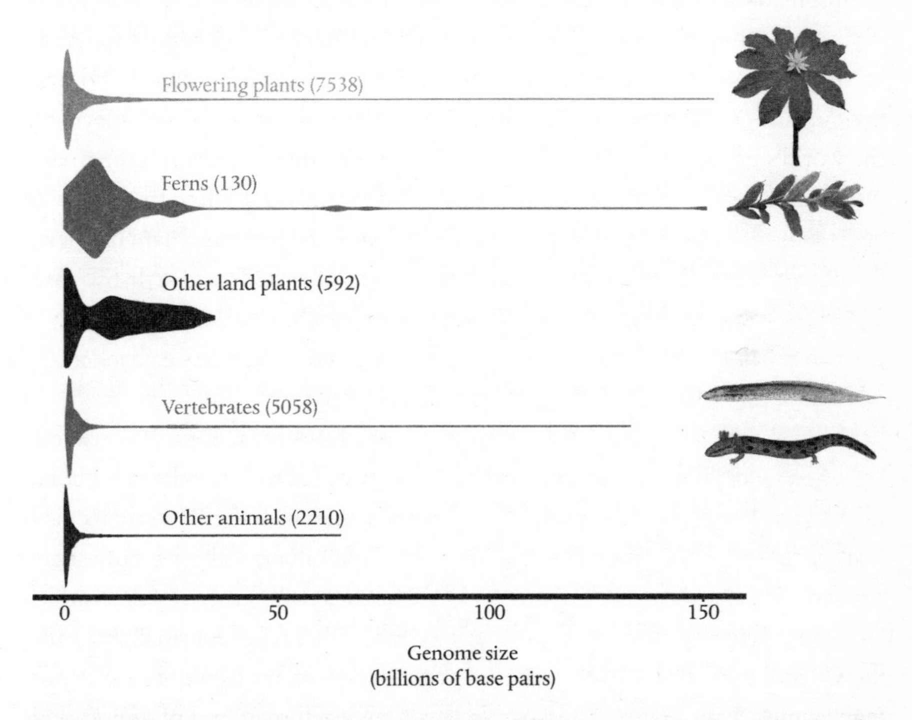


**Figure** 4 The enormous range of genome sizes in different organisms shows that size does not correlate with complexity. The numbers in brackets after each is the number of species analysed. The units of genome size are billions of base pairs.

*Arabidopsis*, a relative of the Brussels sprout, has more genes than either'. The comparison between the genetic content of a chicken and a Nobel laureate is broadly correct; chickens have about one-third the DNA of humans. However, the fact is that all humans, including winners of the 66-mm-diameter gold medals given out at Stockholm, have roughly the same number of genes as the weedy relative of the mustard family, *Arabidopsis*, and poultry (~25 000 genes).

The first plant genome sequence, presented to the world in December 2000, belonged to *Arabidopsis*.[5] The sequences for rice, then a tree, and then a moss, soon followed. Part of the fascination here is that each DNA sequence is really a molecular 'living fossil'. It is an exquisitely detailed archive storing an enormous amount of information about the collections of genes needed to build an organism, and the instructions regulating when, where, and how those genes are expressed. Decode the genomes of individual species, and the past extends an arm to the present to provide unique insights on how they evolved, and bring a new perspective on how they became adapted to their environment. Fascinating and important though these genomic glimpses of the past lives of plants are, genome biology really comes into its own when we compare the sequences of everything from cyanobacteria to trees. Then we discover the ancestry, and changes in the diversity of genes, needed to reconstruct shifts in the genetic make-up of plants during major evolutionary transitions in their history. Called 'comparative genomics', this extraordinary branch of science decodes the archives of fundamental molecular events written into the DNA code millions of years ago and connects them with transformative events throughout the deep evolutionary history of plants.

Obtaining genomic coverage of the plant kingdom to uncover the details of that evolutionary journey is very much a work in progress. We need to sample more of the diversity of species classified across diverse taxonomic groups to avoid the risk of misinterpreting the past.[6] But the sequence data accumulated so far includes taxa on key nodes across the plant tree of life such as chlorophyte and charophyte algae, bryophytes (liverworts, mosses, and hornworts), a lycophyte, and gymnosperms, and is shedding light on the genetic processes shaping the evolution of biological novelty and complexity. Already, the genomic revolution is weighing in on profound questions about the origin, speciation, and extinction of plant life that captivated Wallace and Darwin; not to mention milestones in human civilization, such as the appearance of modern savannas, as the ancestors of humans came down from the trees to live on the grassy plains, and domestication of crops that feed the world.

Prerequisites to being a green plant are an ability to generate energy and an ability to photosynthesize. These functions are served by the mitochondria and chloroplasts of cells that descended from once free-living bacteria that were engulfed by a tiny host algal cell and escaped digestion. Some of the simplest and smallest unicellular marine algae contain a single mitochondrial powerhouse for energy production and a single chloroplast for conducting the business of photosynthesis. They belong to the genus *Ostreococcus*.[7] Celebrated for their diminutive proportions, these tiny spherical algae are classified as chlorophyte algae and form part of the picoplankton living in the world's oceans. Picoplankton form the base of food webs in marine and fresh waters. Roughly the size of a bacterium, these algae are the smallest known free-living eukaryotes (i.e., complex cells with a nucleus). Rightly dubbed by commentators 'the reigning champion of cellular miniaturization',[8] *Ostreococcus* has a correspondingly tiny genome of 12–13 million base pairs (denoted 12–13 Mb) and around 8000 genes.[9] For comparison, the small(ish) genome of *Arabidopsis* is about ten times larger and codes for approximately 27 000 genes.

Those 8000 genes are all *Ostreococcus* needs for running its cell biology and metabolic pathways, including processes like making cell walls, photosynthesis, and respiration. Around 30% of its genes are also present in land plants, with which it last shared a common ancestor over a billion years ago. Of those 30%, land plants use some to control the branching patterns of shoots and the development of leaves. How strange. Obviously, tiny single-celled photosynthesizing algae are not using them for doing such jobs,[10] and these bits of algal DNA are being redeployed for new purposes later down the evolutionary line. There is also the strong possibility that *Ostreococcus* has a special mode of photosynthesis that is able to concentrate carbon dioxide around the photosynthetic enzyme systems to improve efficiency.[11] It represents a simplified version of a mode found in grasses of tropical savannas, the so-called $C_4$ photosynthetic pathway that evolved hundreds of millions of years later. Yet there it is. Complex computational analyses of the *Ostreococcus* genome show that it contains a near-complete set of biochemical reactions for photosynthesizing in this mode (with only a handful of reactions to be added to fill in the gaps).[12] But what conceivable use is this to an alga living in the sea? Probably, the answer is that the pathway enables *Ostreococcus* to cope when algal blooms rapidly deplete dissolved carbon dioxide from the surface waters, causing temporary carbon dioxide starvation.

Plant genomes do not come any more stripped down than that of *Ostreococcus*. Its freshwater cousin, the famous unicellular green alga *Chlamydomonas*, found in stagnant waterbodies, soils, and snow, has a ten-fold bigger genome.[13] It contains nearly three times as many genes as its distant picoplankton cousin, including those required for the usual things organisms do, like swimming, finding food, mating, and building chloroplasts. In evolutionary terms, *Chlamydomonas* diverged from flowering plants about a billion years ago. Compare its genome with that of other organisms and you find it shares about 35% of its genes with flowering plants and humans. It also has an additional 10% it shares with humans, but not flowering plants.

Many of the genes *Chlamydomonas* shares with humans are for building flagella, whip-like structures rotated by molecular motors that function to move it around. Flagella genes have been lost in most seed plants, which evolved novel fertilization mechanisms involving pollen grains that finally freed land plants from their inherited aquatic ancestry and aquatic fertilization.[14] The earlier-evolving groups, such as algae, bryophytes, ferns, and some gymnosperms, produce motile sperm cells propelled along by single or multiple flagella. Animals, including humans, retained and modified these useful structures in a shortened form to make fine, waving threads called cilia. Cilia play crucially important roles in the normal functioning of our brains, our kidneys, where they bend as fluid passes over them allowing cells to sense flow, and our lungs, where they beat in co-ordinated waves to sweep up dirt and mucus.[15] Signs of our single-celled ancestry spring from every cell in our bodies. If we develop faults in ciliary proteins, our health can suffer. Defects in cilia are linked to many illnesses, with one of the best-documented examples being a rare human genetic disease called Bardet–Biedl syndrome, which affects anatomical development, leading to extra fingers, blindness, and obesity. The identification of one of the human genes linked to the disease (*BBS5*), that disrupts the making of cilia, came from analysis of the genome of the humble alga *Chlamydomonas*.[16]

Being green and single-celled is a promising start, but what the history of life on Earth makes clear, for plants at least, is that eventually a multicellular lifestyle beckons. Once life gets going, the emergence of complex multicellular life forms from single-celled ancestors is seemingly inevitable.[17] This momentous event probably happened around 700 million years ago when single cells clubbed together, perhaps using flagella and other similar structures, to co-operate, communicate, and form the first multicellular green algae. What did it take to make the crucial evolutionary transition from single-celled to multicelled algae? The question is a profound one, as the arrival of multicellular green algae in the late Precambrian ocean proved a watershed moment in Earth's history, presaging the invasion of the land by millions of years.

The answer is that multicellular plants required genetic toolkits for solving a new set of challenges, such as how to get cells to adhere and communicate with each other, and how to get oxygen into tissues as the organism becomes larger. Co-operation between individual cells is the key, because in a multicellular organism, the fates of the many cells that make up the new entity become bound collectively to the fate of a single individual. Decoding the genome of the beautiful multicellular green algae *Volvox* provided the best insights for how this might have happened. Around half a millimetre in diameter, *Volvox* is just visible to the naked eye. Antonie van Leeuwenhoek (1632–1723), the Dutch father of microscopy, first viewed these little round green 'animalcules' with utter fascination over three centuries ago.[18] Using microscopes of his own design that could magnify by 270 times, he observed *Volvox* in drops of pond water with a motion that 'twas wonderful to see'. Later named *Volvox* by Linnaeus in 1758 (from the Latin *volvere*, to roll), these delightful green 'fierce rollers' move through the water by division of labour into two cell types—small and large. Each green sphere consists of some 2000 of the small cells (the number varies with species), embedded into a gelatinous matrix on the outer surface, which use pairs of flagella beating in waves to roll the colony forwards. Some cells form anterior eyespots, enabling the colony to swim towards light. Inside are 16 or so much larger sex cells involved in reproduction.[19]

Surprisingly, for all its multicellular glory, the genomic distinction between *Volvox*, with its complex life cycle, and *Chlamydomonas* is almost trivial and did not apparently involve large changes in its genetic repertoire.[20] Both organisms have around 15 000 genes, with few of those in *Volvox* being different from *Chlamydomonas*, though of course, we find extra genes involved in building that gelatinous matrix—the glue that holds the many cells together.[21] The evolutionary transition from being single-celled to multicellular is not large increases in genes but changing the way those genes are *used*. Back in the Triassic, 200 million years ago, it took *Volvox*'s predecessors some 35 million years of tinkering with the ancestral algal genetic blueprint to navigate towards the transition from single cells to multicellular colonies.[22]

The key point to remember here is that algae such as *Chlamydomonas* and *Volvox* belong to the great chlorophyte division of algae that is distantly related to land plants. The truth is that land plants evolved from freshwater charophyte algae, from which they inherited an array of biochemical, physiological, and cell biology attributes. We can expect, then, that the genomes of charophytes hold deeper genetic insights into the origin of those traits that allowed algae to give rise to terrestrial plant life. At least that is the standard line of thinking. Yet, various complex lines of evidence are leading to the growing suspicion that charophyte algal lineages may have evolved to occupy terrestrial environments before giving rise to land plants.[23] The idea that freshwater algae were the first 'invaders of the land' gives you pause for thought. If this scenario is correct, we can anticipate 'terrestrialization' signatures for conquering soil and sediments to be written in their genomes. And, to some extent, this view was confirmed when a team sequenced and analysed the genome of a filamentous terrestrial alga, *Klebsormidium flaccidum.*[24] *Klebsormidium* has the genetic machinery for protecting its photosynthetic apparatus against damage from high intensity light, typically encountered on land, and for synthesizing hormones and metabolic signalling pathways to survive environmental stress.

Not only that, but it was further reinforced when the genome biologist Stefan Rensing undertook the most comprehensive genome sequencing project yet of a charophyte alga, focusing on a species that lives fully in freshwater called *Chara braunii. C. braunii* has a wonderfully elaborate body plan and is more closely related to land plants than *Klebsormidium.* Its genome codes for ca. 23 500 genes,[25] substantially more than *Chlamydomonas* (ca. 19 000) or *Klebsormidum* (ca. 16 000), but similar to some land plants (e.g. moss ~35 000 and *Arabidopsis* 27 000). Much of the expansion of gene numbers in the genome of this species underpins its remarkably complex morphology, including its rhizoids and elaborate structures for sexual reproduction. The genome also has both striking similarities to land plants, and important differences. Rensing's results are redefining the list of whole families of genes that were thought previously to be specific to land plants and, we now realize, are important for living in shallow freshwater environments that occasionally subject the plants to dry air.

Sequencing the genomes of additional charophyte species is an urgent priority, but in the meantime others have adopted the next best thing, that of determining a compact 'snapshot' of a species known as the transcriptome. Obtaining such a snapshot is a neat trick of cell biology pioneered by scientists from the human genome project. It involves isolating and extracting the messenger RNA from the tissues of the selected organism. Messenger RNA is present inside all living cells and is synthesized during the first step in gene expression to mirror the DNA code. These stringy molecules are exported out of the nucleus carrying the amino acid code for making proteins. Complex molecular machines (ribosomes), which read the message and link the specified amino acid building blocks together, carry out the conversion of messenger RNA into proteins. But the step that interests us here is the conversion of the original DNA code into the complementary language of messenger RNA molecules. In the lab, an enzyme called 'reverse transcriptase' is used to translate these snippets of RNA back into the original DNA message. The sequencing of the DNA message, essentially a copy of a gene, is then undertaken to determine the identity and order of the bases. What you end up with is a condensed version of an organism's genome containing all genes that are being expressed (i.e., those that are active) in the tissues sampled at a given time—the transcriptome.

Obtained in this way, the transcriptomes of charophytes confirm that these diverse algae have sophisticated collections of genes responsible for the synthesis, perception, and signalling of major plant hormones.[26] As anticipated, they also have a genetic repertoire predisposing them to make the bold transition from water to the land—a repertoire absent from the chlorophyte algae. The transcriptomes of the filamentous *Spirogyra* (Zygnematales) and the flying-saucer-shaped *Coleochaete* (Coleochaetales)[27] have been analysed in detail, and found to harbour the genetic machinery for two important cellular features, plasmodesmata and phragmoplasts. Plasmodesmata are the pores in plant cell walls that allow neighbouring cells to communicate with each other and distribute water and nutrients to nourish the developing embryo. Phragmoplasts are microfibres involved in cell division that these algae share with all land plants. The early ancestors of land plants underwent cell division without the help of phragmoplasts but the course of evolution gradually modified the cell division apparatus to provide a higher degree of developmental flexibility.[28] By flexibility, I mean it allowed asymmetrical cell division in shoot tips and/or rotation of the plane of cell division to permit branching. In other words, it helps provide the fundamental basis for 3D plant development.

The earliest proper land plants that descended from these algae were tiny, rudimentary photosynthetic organisms. They left behind mysterious fossilized spores in the rock record as tell-tale signs of their existence, and little else. The decoded genomes of the bryophytes (liverworts, hornworts, and mosses), the oldest surviving lineages of those pioneering plants, speak to us of the genetic innovations that define a terrestrial plant. Bryophytes entered the genomic era back in 2008, when the DNA sequence of a moss (*Physcomitrella patens*) was published in the journal *Science*, by a team also led by Rensing as it happens.[29] The sequence of a liverwort (*Marchantia polymorpha*) followed a decade later,[30] and the first genome sequencing of a hornwort (*Anthoceros agrestis*) is in the pipeline.[31] These genomes are giving us clues as to what it took to make a multicellular terrestrial plant from a charophyte algae ancestor.

Consider the case of the liverwort genome, which hints at some of the genetic innovations that define land plants. Liverworts are among the earliest diverging lineages of land plants and grow as flat, green photosynthetic cushions. They have probably lost some ancestral traits and acquired a mosaic of others. Other groups of liverworts include 'leafy' liverworts that have a more familiar morphology, with a stem and leaf-like extensions. The *Marchantia* genome has ca. 19 000 genes, a few thousand less than *Chara*. Some of these genes are shared with these green algae, others with vascular plants to the exclusion of green algae, and still others are uniquely its own. An expansion of the diversity of genes regulating the genome has been detected and these correspond with an increased diversification of developmental and environmental responses compared to green algae. Presumably these are useful for coping with the wider range of environmental conditions found on land compared to life in the water. The genome also has components of the flowering plant-like machinery for plant hormones, genes encoding 'novel' biochemical pathways to cope with the stress of bright sunlight, and components for cuticle and wax biosynthesis to prevent water loss and protect against insects. These genomic attributes are again consistent with plants gaining critical features of their biology during the evolutionary transition to land. There is also strong genomic evidence of substantial gene transfer from fungi and bacteria, including genes involved in the biosynthesis of chemical compounds called terpenes that contribute to oil bodies, defensive organelles unique to liverworts.

One lesson to be drawn from analysing the genomes of bryophytes is that duplicating and tinkering with bits of DNA inherited from algal ancestors can generate the evolutionary novelty needed to build and operate a multicellular land plant. Genes of algae, it seems, can usefully serve land plants with some evolutionary fiddling in how they are wired together. There is an important general principle here: evolution works with what is to hand. As we have already seen, some of the genetic material for the physiological adaptations needed for living on land was likely already present in the algal ancestral lineage, with much of the extra genetic material needed for developmental innovations originating in the ancestral land plant. We see the pattern repeated when vascular plants, with their hallmark internal plumbing connecting roots to leaves, originated.

Lycophytes (clubmosses, spikemosses, and quillworts) are the descendants of early-evolving vascular plants and represent an interim step in evolution, reproducing by spores instead of seeds. Decoding the genome of the spikemoss *Selaginella moellendorffii* (Selaginellaceae)[32] was the first step undertaken by a large consortium of scientists. Comparing it with the genomes of algae, moss, and flowering plants was the next step. The results showed it took a trifling number of extra genes (about 500) to make a vascular land plant, as represented by spikemoss. The later evolution of a flowering plant required an extra 1350 genes. Put another way, three times as many genes were required to invent a flowering plant starting from a lycophyte baseline than were required to invent a vascular plant starting with a bryophyte baseline.[33] At least this is true for a spikemoss: coverage of species within particular lineages, whether algae, bryophytes, or lycophytes, is extremely limited in terms of genome sequencing, so we should be cautious about drawing firm conclusions regarding the importance of changes in gene numbers. More important is that these examples serve as model plant species and this represents one of the great successes in modern biology, allowing us to investigate a wide range of long-standing biological questions about evolutionary mechanisms and processes.[34] We can illustrate the point by looking at how plants 'invented' vascular tissues for transporting water and the lignin needed to strengthen these specialized water-conducting cells.

As we might expect, some of the extra genes found in the spikemoss genome compared to the moss and liverwort genomes are involved in synthesizing vascular tissues. These tissues are comprised of specialized cells for transporting water (xylem) and sugars (phloem). Mosses lack xylem and phloem but do have bundles of simple hollow water-conducting cells and phloem-like cells whose development is genetically controlled. Genes carrying the detailed instructions for vascular plants to make hollow, water-conducting xylem cells were discovered to be closely related to those used by mosses for the formation of their conducting bundles.[35] The liverwort *Marchantia*, which has no internal water-conducting cells, has a rudimentary network of the same genes. Charophytes possess a handful of related genes whose function is uncertain. All of which suggests something extraordinary. Vascular plants 'learned' to make their vascular tissue from ancestral pre-vascular land plants.

The appearance of vascular tissue presented its own challenges because these specialized hollow cells, linked end-to-end, needed stiffening to prevent collapse and support the stem. An ability to synthesize the immensely strong and recalcitrant polymer lignin provided a solution. Lignin gave plants structural rigidity and the ability to stand upright. Lignin explains why trees are woody and why wood takes so long to decay. The extensive coal forests that developed throughout the Carboniferous and Permian (ca. 323–252 million years ago) were composed of giant arborescent relatives of clubmosses, as well as lignin-producing ferns, horsetails, woody shrubs, and trees related to conifers. During burial in sediments, the organic remains of the plants are slowly transformed into peat, lignite, and then coal. Massive coal deposits formed in this way developed largely because of widespread waterlogged conditions inhibiting decay, thanks to a combination of climate and tectonics during the formation of the supercontinent Pangaea. Towards the end of the Permo-Carboniferous, as the interval is known, burial of organic matter slowed as the climate dried out, drawing to a close an extraordinary era of coal formation.[36] Hundreds of millions of years later, that coal was to fuel our own industrial era. That, at least, is the widely accepted standard view.

A complementary hypothesis emerged when David Hibbett of Clark University sequenced the genomes of a wide variety of fungi and used this information to reconstruct their history, and specifically when white rot fungi gained the metabolic capacity to degrade lignin and unlock its concentrated carbon energy. Hibbett's group suggested that a diverse and versatile arsenal of lignin-digesting enzyme pathways, comparable to those of modern white rot fungi, may have arisen late in the Permo-Carboniferous.[37] Estimates of the range of dates are sensitive to the choice of fossils used to calibrate the molecular clock involved and have considerable uncertainty. Granted these uncertainties, Hibbett suggested that after they had evolved, they assisted decomposition of the lignin-rich woody debris of early forests and slowed coal formation.[38] Intriguing fossils resembling modern white rot occur in seed ferns from the Permian (260 million years ago), and gymnosperms from the Triassic (230 million years ago), consistent with the inferences from genomics that these fungi were on land at around the right time. But not everyone is convinced by this elegant line of reasoning, and the sceptics firmly reject claims that delayed fungal evolution played a role drawing to a close peak coal formation during the Palaeozoic.[39] In their view, coal accumulation patterns reflect the unique combination of climate and tectonics and nothing else. The debate highlights the challenges facing those looking to integrate a genomic perspective into Earth's complex history.

The invention of lignin probably begins with the first land plants, not algae.[40] To be clear though, the biosynthetic machinery for making chemical compounds required to help multicellular plants cope with the challenges of a terrestrial existence is evolutionarily very ancient.[41] The point is that occasional reports of discovery of lignin in charophyte algae have proved controversial and not yet withstood further scrutiny. The moss *Physcomitrella* has a handful of genes for synthesizing the basic chemical scaffold of the stuff,[42] whereas our freshwater green alga *Chlamydomonas* has none. The chemical complexity of lignin is such that even a scaffold has its uses for moss because it absorbs wavelengths of solar energy in the ultraviolet range, protecting emerging land plants. Components of the biochemical apparatus involved in its synthesis were likely present in ancestral freshwater algae, which used them to make the tough protective outer coating of spores—sporopollenin.[43] The spikemoss genome includes genes enabling it to synthesize proper lignin to give its water-conducting cells structural rigidity. The steps by which the pathway for the synthesis of lignin was assembled are not yet known; some of the genes may even have originated from symbiotic fungi and bacteria.[44] Other lineages of vascular plants, including ferns, gymnosperms,[45] and flowering plants,[46] invented the same type of lignin, but not always via the same biosynthetic pathways. Lignin is a rather wonderful molecular example of convergent evolution—the convergence of different evolutionary lineages on the same, or a very similar, solution to a problem. Think of the independent evolution of wings for flight in bats and birds, and echolocation in dolphins and bats, which are classic examples from the animal kingdom.[47]

Lignin also makes trees possible. Once plants mastered the ability to synthesize lignin, it was not too long, in an evolutionary sense, before trees evolved, as the rich Devonian record of fossil plant discoveries makes clear. In fact, the molecular basis for differences between woody perennials and annuals appears to be rather small. Suppress the activity of only a couple of genes in *Arabidopsis*, for example, and you can extend its growth phase and make it produce wood.[48] We can speculate that it simply was not that difficult for plant life to evolve perennial growth and the woody habit characteristic of trees. Indeed the tree habit has arisen independently multiple times throughout plant evolution—eventually the monocotyledons got in on the act (think of dragon trees and palms).

Another line of reasoning supporting this view comes from what happens on islands when plants have free rein, as both Charles Darwin and Alfred Wallace were well aware. They were among the first naturalists to report that, given the chance, island-dwelling herbs soon exhibit a tendency towards a tree-like growth habit. Astonishing evolutionary events on the Macaronesian islands, located in the Atlantic between Europe and North Africa, have been unravelled by analysing the DNA of plants belonging to the genus *Sonchus*, a member of the daisy family (Asteraceae). According to the DNA evidence, several species of tree on the islands, including *Sonchus canariensis*, evolved from the small, closely related herbaceous ancestral plant *Sonchus asper*, within a few million years of its appearance on Gran Canaria or Tenerife in the Canary Islands.[49] One hypothesis is that extinctions in the Canaries following the Pliocene glaciation 2–3 million years ago created open habitats into which tree-like forms could evolve and diversify.

It is a similar story for those botanical glories on the desolate St Helena, an isolated speck of a volcanic island where the fallen emperor Napoleon was exiled for the rest of his life over 200 years ago: 'gumwood' (*Commidendrum*) and 'cabbage' (*Melanodendron*) trees.[50] All species of *Commidendrum* and *Melanodendron* are threatened, and officially recognized as such by the International Union for Nature Conservation. They survive today as small relict populations on cliffs or in fragmented patches of vegetation. As in the Macaronesian islands, the DNA of these forest trees, which also belong to the Asteraceae, indicates they evolved from a single common shrubby ancestor that arrived on the island, probably from mainland South Africa. By fortuitously arriving on St Helena, these plants found the niche of forest trees unoccupied and soon exploited it, evolving into diverse species of proper trees with trunks, spreading crowns, and woody rooting systems. Trees are obviously admirable opportunists, and the staggering estimated three trillion trees in forests worldwide today constitute over 80% of the biomass on the continents, and harbour 50% of terrestrial biodiversity.[51]

As they support, shelter, or feed an enormous diversity of animals and microbes, it's no surprise that the genomes of trees such as the black cottonwood (*Populus*),[52] and the common fast-growing Australian *Eucalyptus grandis*,[53] have large numbers of genes deployed for doing what a tree needs to do: fighting attacks from harmful organisms like fungi, bacteria, and insects. Protecting themselves from the relentless onslaught of attack by pests and pathogens requires synthesis of chemical defence compounds. *Eucalyptus*, with its oil-containing leaves, releases numerous volatile compounds, synthesized by a large collection of more than 100 genes, for defence against a wide array of pests and pathogens; a moss, for comparison, has just two, and a liverwort, only a few more. The genome of *Eucalyptus* also carries the signature of the unusual reproductive biology of this tree: its unopened flower buds are covered with a pixie hat, a strange feature reflected in its name—*Eucalyptus* derives from the Greek *eu-*, meaning well, and *kaluptos*, meaning covered. During the blooming process, the pixie hat falls off the vase-shaped floral buds, and the flowers that develop lack visible petals. Inside the genome, we find that genes regulating the formation of elaborate flowers have gone, while genes for producing large amounts of pollen and seed have massively increased compared to other flowering plants.

A recurring feature of the various genomes we have been considering is a general trend of increasing gene numbers more or less along a gradient of increasing plant complexity moving from algae to trees.[54] As if by magic, gene numbers handily increase, and plants are free once again to systematically advance in organized complexity. Where do all of these extra genes originate? Whole genome duplications can provide one major source. During normal cell division, the parent cell divides into two or more daughter cells, with each cell receiving the normal complement of chromosomes. Failure in the normal process of cell division and chromosome duplication can lead to a genome doubling, with each chromosome having a twin. The process effectively creates a 'spare' genome, potentially enabling thousands of extra gene copies to be free to evolve new functions. Genome doubling can also occur when the merging of chromosome sets of closely related species follows failure in cell division. This is the basis of hybridization. Organisms (plants or animals) possessing more than two complete sets of chromosomes are known as polyploids and the scientific investigation of polyploids dates back over a century,[55] to work by the distinguished botanists Hugo de Vries (1848–1935) and G. Ledyard Stebbins (1906–2000). The formation of polyploids is a generally accepted process of speciation, the evolution of new species.[56]

Only with the advent of new technologies in the era of genome biology did scientists begin to document whole genome duplication events that were tens of millions or hundreds of millions of years old. Detecting the remnants of ancient genome duplications is a high art form, with the search for traces of genetic events that happened millions of years ago demanding computational approaches to identify blocks of duplicated genes and the application of different techniques for dating the duplication events.[57] Few ancient genome duplication events survive, suggesting that they are often evolutionary dead ends, leading to genomic shock and sterility. Occasionally, though, duplications can have a profound impact on evolution by providing two copies of each gene for evolutionary forces to work on.[58] We shall learn how shortly, but for now we can marvel at the theoreticians who delight in analysing the genome of *Arabidopsis*. It is worth unpacking a little of what they have found so far because it tells us about the fate of all those extra genes. *Arabidopsis* shows evidence for having undergone three whole genome duplication events of varying antiquity over the past 150 million years. Scientists have found that the first doubling created 17 193 duplicate genes with only ~4% retained; the second doubling created 20 316 duplicate genes with ~14% retained; and the third doubling created 24 351 duplicate genes with ~16% retained.[59]

The lesson from these numbers is obvious. Whole genome duplications are not genetic stockpiling events, far from it. Most duplicated genes are quickly erased from the genome on a 'use it or lose it' basis. Expansion of gene numbers happens with the gradual accumulation of small numbers of surviving genes over time and frequent duplications of individual genes themselves. Gene survival is highly selective, and regulatory genes, which influence the expression of other genes, are retained in preference to genes regulating metabolism. In fact, when a whole genome duplicates, all of the genetic material is replicated, not just the genes, and plants have mechanisms for eliminating chunks of unnecessary DNA with varying degrees of efficiency. For example, the genome of the bladderwort, *Utricularia gibba*, a small carnivorous plant that lives in freshwater and damp soils, is compact, despite having undergone three rounds of whole genome duplication. The efficient removal of large amounts of repetitive DNA has compressed the genome, leaving the remainder to encode for roughly the same number of genes as *Arabidopsis*.[60]

At the opposite end of the spectrum sits the huge genome of the distinctive gymnosperm tree *Ginkgo*, with its beautiful heart-shaped leaves.[61] *Ginkgo*'s genome is 80 times larger than that of *Arabidopsis*. Some of its 40 000 genes may explain *Ginkgo*'s extraordinary resilience to insects. It has genes for synthesizing chemicals that fight insect attack directly and others for attracting the enemies of plant-eating insects by synthesizing and releasing volatile organic compounds. The *Ginkgo* genome is bigger than the notoriously large maize genome, but only half the size of the enormous genome of the Norway spruce, *Picea abies*, which has been inflated by the accumulation of large numbers of repetitive mobile elements. The huge size of the *Ginkgo* genome also reflects a very high proportion (more than 75%) of repetitive sequences, resulting from both gradual accumulation over deep time, and from two ancient whole genome duplication events.

Whole genome duplications in the distant past gained a new significance when Claude dePamphilis and his team published a landmark investigation into ancient whole genome duplications in plants in the journal *Nature*.[62] They concluded that all extant flowering plants shared a very old whole genome duplication (dating to around 234 million years ago) and that all extant seed plants shared an even older one (dating to approximately 349 million years ago). In other words, the appearance of seed plants and the origin of flowering plants are coincident with whole genome duplications (**Figure** 5). Both duplications are marked by 'sudden' bursts of gene proliferation, with thousands of new genes appearing simultaneously. These surviving genes may have formed new genetic circuitry to regulate new developmental processes important for building the major innovations like seeds and flowers that contributed to the rise of seed plants and flowering plants. But here is the rub. Not everyone is convinced by this apparently elegant story. In the best traditions of science, a different team re-analysed the data sets and felt it was premature to reach a definitive conclusion on the exact number and timing of ancient duplications.[63]

As ever, there is a need for more high-quality data on additional flowering plant and gymnosperm species, and better computer algorithms, to understand better

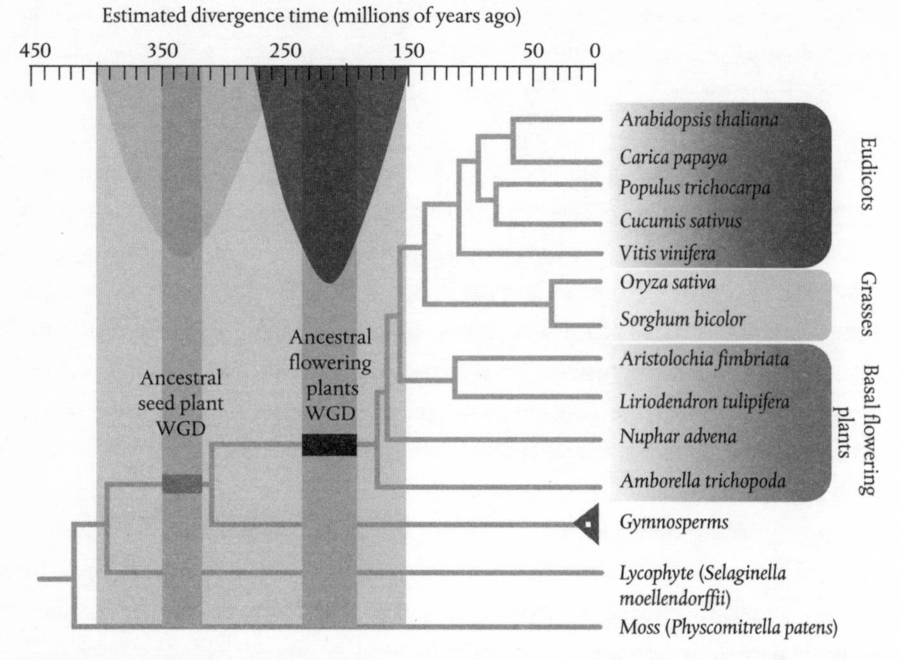


**Figure** 5 Controversial evidence suggesting whole genome duplication (WGD) events in the ancestors of seed plants and flowering plants. Eudicots are a clade of flowering plants.

the genomic driving force of plant evolution. Nevertheless, the fingerprints of gene duplication in driving innovation are found in the details of flower evolution. Enrico Coen and Elliot Meyerowitz proposed the elegant ABC genetic model of floral identity over twenty-five years ago.[64] In their model, the action of several genes together specify the formation of the component parts of flowers—sepals, petals, stamens, and carpels (**Figure 6**). Moving from the outside of the flower inwards, the sequence of gene actions is as follows. The A function genes specify sepals, the A and B function genes specify petals, B and C function genes specify stamens, and C function alone specifies carpels. Finally, the E function genes are required in conjunction with those of the other floral regulators for correct organ specification.

Today's flowering plants all share more or less the same core ABC genetic circuitry to produce the enormously wide variations in flowers, and many of the genes controlling the different functions are duplicates of each other.[65] The startling implication is that the earliest flowering plants possessed a full complement

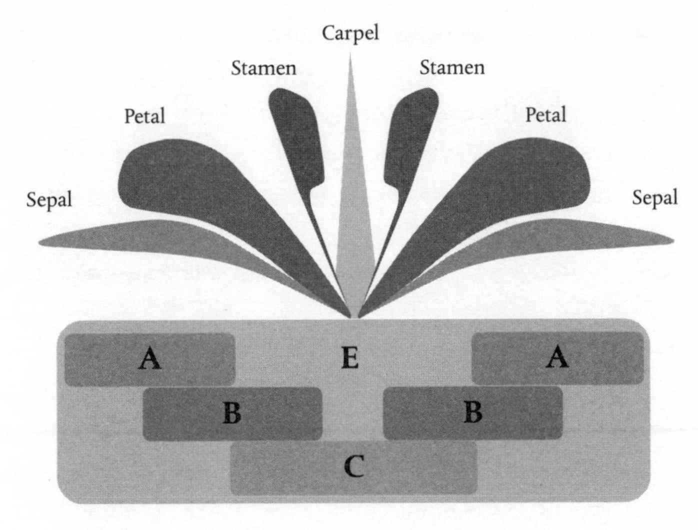


**Figure 6** The ABC genetic model of flower evolution.

of the basic genetic information needed to assemble a modern flower. Increasingly elaborate flowers may have evolved by refashioning the developmental circuitry that originally controlled sterile leaves. In fact the origins of flowers go further back still because a similar gene system already seems to have existed in the ancestral seed plant, given what we know about the role of B and C class genes in gymnosperms. C class genes specify a female cone, whereas B plus C classes specify male cones. Creating a flower, then, requires the evolutionary assembly of these 'cones' in the same shoot tip (rather than separate organs as in gymnosperms) to create a bisexual shoot, adding some more sterile whorls surrounding the sexual organs with A class genes, and finally enclosing the ovules with fused leaf-like organs, i.e., a carpel.[66]

Insights from genome biology go further by offering an original take on Darwin's 'abominable mystery'—what he perceived to be the abrupt origin and dramatic radiation of flowering plants. In Darwin's time, flowering plants were thought to enter the fossil record very suddenly and apparently 'fully evolved' in the mid-Cretaceous, about 100 million years ago. This prompted him to speculate about a long pre-Cretaceous history for flowering plants, perhaps in remote areas that left no fossil signals for palaeobotanists to unearth. With the benefits of continued exploration and discovery of the fossil record since Darwin's day, the date of the first flowering plants has been pushed back to the early Cretaceous, 140 million years ago.[67] In a sense, then, Darwin's mystery is solved. Flowering plants originated far earlier than he thought and flowers, as we have seen, arose partly because genome duplications provided the genetic material for flower evolution. But how do we explain their explosive diversification within a few tens of millions years in the Cretaceous?

One answer, provided by genome biology, links diversification to duplication. Whole-genome duplications have been reported for the Brassicaceae (3700 species), Asteraceae (25 000 species), Fabaceae (19 400 species), and the Solanaceae (over 3000 species), to name but a few.[68] In these families, genome duplication seems to correlate with species-rich plant families. The common ancestor of the Asteraceae (daisies), the largest family of flowering plants, for example, has a well-dated genome duplication event tightly linked to sharp increases in diversification rates.[69] The origin of the monocotyledons, which gave rise to some 11 000 species of grasses, is also associated with a whole genome duplication event.[70] The origin of grasses with the $C_4$ photosynthetic pathway in tropical savannas, the expansion of which at the expense of low-latitude forests took place during the Miocene (11–5 million years ago), involved grasses of the Andropogoneae tribe. Grass species in this tribe experienced more than 30 genome duplications in a comparatively narrow interval of geological time, tempting us to speculate that this helped them deal with the changing climate and atmospheric conditions of the time by giving them new ecological tolerances.[71]

Later, from about 10 000 years ago, domestication and cultivation of wild grasses began in the Fertile Crescent of the Middle East, and saw hunter-gatherer societies replaced by farming communities. Five grasses—rice, maize, wheat, barley, and sorghum—emerged from the process, and today feed the world. The ancestor of these grasses underwent a whole genome duplication event around 90 million years ago and again 70–50 million years ago[72] to create the ancestral genome found in modern cereals. The great insight leading to this picture came from the simple observation that the order of blocks of genes on the chromosomes of grasses is the same. It is the same in rice and wheat, for instance, despite hugely different genome sizes. Clever detective work showed how each cereal genome derives from the cleavage of a single structure, the hypothetical 'ancestral' genome.[73] The genomes of maize, sorghum, and rice record their millennial dialogue with humans during domestication, telling us how farmers on the American, African, and Asian continents all independently selected for similar traits—higher harvests and greater resistance to disease. The story is written in the genomes of the cereals we consume without a second thought.

We can also tie specific innovations in flowering plants to genome duplication events that ignited an arms race between plants and insects and contributed to the explosive diversification that puzzled Darwin. Consider the brassicas, the genus that includes mustards and cabbages, which defend themselves against herbivores by synthesizing protective chemical compounds. Nearly 90 million years ago, the ancestors of brassicas developed a particular set of chemical defence compounds (called glucosinolates) that are toxic to most insects, and as a consequence triggered an evolutionary arms race with the caterpillars of the butterflies that fed on the plants.[74] Each time the brassicas evolved increases in chemical defence complexity, a burst of plant diversification followed. In response, the brassica-feeding butterflies (Pierinae) evolved the ability to detoxify the compounds. Freed from the toxicity of their food plants, butterflies underwent their own burst of diversification until the plants escalated the situation by evolving another chemical elaboration to defend themselves.

Modern DNA sequencing has shown genome duplication events enabled advances in the chemical weapon sophistication of the brassicas, with the arms race between plants and butterflies that started 90 million years ago continuing; modern brassicas are now able to synthesize more than 120 different varieties of glucosinolates. The biologists Paul Ehrlich and Peter H. Raven first proposed the idea that plant–insect interactions generate biodiversity nearly half a century ago.[75] Written in the DNA code, we later discovered, is evidence supporting their hypothesis, with repeated escalation between chemical innovations and bursts of diversification in brassicas and Pierinae butterflies. The parsley family (Apiaceae) and the butterflies (Papilionidae) and moths (Oecophoridae) that attack it are probably also engaged in a similar arms race; the details are yet to be deciphered. None of this is to deny the traditional co-evolution of flowers and their pollinators, or fruits and their dispersers, as drivers of biological diversification, but rather to show how genome biology offers fresh explanatory insights into modern biological diversity.

**Figure** 7 The Korean-born geneticist Susumu Ohno (1928–2000).

Over half a century ago, Susumu Ohno (1928–2000) (**Figure** 7), a prominent and influential biologist, proposed an extraordinary idea. In his 1970 book, *Evolution by Gene Duplication*, he presciently postulated that two rounds of whole genome duplication accompanied great transitions in the evolutionary history of life on Earth.[76] The mainstay of evolutionary theory (at that time) was that random mutations in existing genes can lead to organisms being better adapted to their circumstances, thereby making them more likely to survive and reproduce. In this way, natural selection could gradually improve the 'fitness' of a lineage in the long-run. But in the preface to his prescient book, he wrote:

> Had evolution been entirely dependent upon natural selection, from a bacterium only numerous forms of bacteria would have emerged. The creation of metazoans, vertebrates, and finally mammals from unicellular organisms would have been quite impossible, for such big leaps in evolution required the creation of new gene loci with previously nonexistent function.

The basic idea is that the extra gene copies resulting from a genome duplication might, in the right circumstances, evolve new functions, for major innovations in body plans as well as other modifications, whilst allowing the functions of the original genes to be maintained. Ohno's radical idea is distinguished from what went before because he proposed evolution through the duplication of genomes, rather than alteration to individual genes. He insisted organisms could deal with extra genomes more readily than they could duplicate copies of individual genes. His idea was that changes in proteins coded for by an extra copy of a particular gene could upset the metabolism of an organism more than the duplication of the entire genome. The explanation is down to a phenomenon called dosage compensation, a process by which organisms equalize the expression of genes to compensate for the change in the normal number of chromosomes present.

By proposing this explanation, he was resurrecting an earlier idea originally advanced by the evolutionary biologist J.B.S Haldane (1892–1964). In what became known as his 2R hypothesis (R for Round of duplication), Ohno suggested a first round of duplication made the transition from invertebrates to vertebrates possible, and a second led to the diversification of vertebrates. Of course, the naysayers were understandably critical right from the off, given the paucity of data at that time. Indeed, back in the 1970s, the idea was widely regarded as 'outrageous', not least because of the lack of sufficient genomic information to test it. With the explosion in modern biology of molecular genetics and genome sequencing data, Ohno's radical suggestion could be critically evaluated, and the evidence suggests he may have been right all along.

Over the course of 500–600 million years of vertebrate evolution, two rounds of whole genome duplication have since been detected, both near the base of the vertebrate tree of life, and both followed by substantial increases in complexity.[77] Rapid innovation, leading to enhanced nervous, endocrine, and circulatory systems, enhanced sensory organs, complex brains, skulls, vertebrae, an endoskeleton, and teeth, followed the first duplication deep in the Cambrian.[78] These changes were followed in vertebrate lineages by innovations such as paired limbs/fins, hinged jaws, and improved immune systems. Later, bony fishes (teleosts) were found to share a third rapid diversification dated to 316–226 million years ago, straddling the end-Permian mass extinction; this fits with Ohno's proposed third duplication, which he added later in his career. Long before all of this evolutionary action, an exceptionally ancient gene duplication may have happened before plants split from animals and fungi[79] (~2.7 billion years ago). Today, it seems more than a coincidence that whole genome duplications are followed by substantial increases in the complexity of major groups of organisms throughout the history of life on Earth, including plants, fish, vertebrates, and fungi. Not even Ohno foresaw this radical possibility. Towards the latter part of his career, he famously married chemical and musical composition, turning DNA sequences into musical pieces. To the delight of many (others were appalled), he had the results recorded by a violinist and a pianist, turning the genetic code of primitive animals into passages of sound.[80]

As the genomic revolution progresses, we are discovering that most species of animals and plants are descended from ancestors that underwent whole genome duplications. Genomics shows how rare but important duplications are tied to innovation and diversification of plants at critical moments in their evolutionary history. The genomes of flowering plants show a wave of duplications dating to between 75 and 55 million years ago.[81] Precise dating of the event is uncertain but they all cluster suspiciously close to the end-Cretaceous mass extinction that saw off the dinosaurs and many other species, 66 million years ago.[82] It was a time of catastrophic environmental change caused by the double whammy of a bolide impact near Chicxulub, Mexico, and periods of massive volcanism, recorded in the form of the Deccan Traps in India. In the face of such environmental instability, did whole genome duplications help plants survive by conferring them with broader environmental tolerances? Intriguingly, conifer genomes also record evidence for an older duplication event, dating back to a broad window of geological time spanning the even more devastating Permian–Triassic extinction, 251 million years ago.[83] Mutated fossil conifer pollen has been discovered in rocks dating to the end-Permian, when a prolonged bout of intense UV radiation pierced a damaged ozone layer drastically thinned by noxious gases emitted during the massive Siberian volcanic eruptions of the time.[84] Pollen is essentially a male sex cell (gamete) and, hit with high doses of UV radiation, it may not have undergone the normal reduction of its chromosome complement. Such a mutational bonanza may have helped conifers survive the catastrophic end-Permian environmental deterioration caused by the Siberian volcanism.

These speculative scenarios raise the question: why would plants that underwent whole genome duplications be more likely to survive drastic global environmental upheaval linked to mass extinctions? One explanation revisits our earlier theme—generating novel gene functions by creating spare copies of genes for evolution to work with. By increasing variation in gene expression patterns, and rewiring some genetic networks of their developmental circuitry, plants have greater potential for tolerating a wider range of environmental conditions than their progenitors with a narrower gene base. In other words, duplicated genes increase the organism's chances of survival under new or harsh environmental conditions. If plants gained an extra genome by merging chromosome sets from a closely related species, then novel combinations of genes might broaden their ecological tolerance. We are dealing with the phenomenon of 'hybrid vigour'. Hybrid plants tend to be better at out-competing species that did not undergo genome duplications because they generally have greater stress tolerance, faster growth, and greater resistance to pests and diseases. No wonder hybrids are frequently invasive species colonizing new or disturbed habitats.

If extraordinary claims for genome duplication events hold up, and the evidence is mounting, we can tentatively conclude that genomic upheavals in the past prompted the evolution of new plant life from ancestral forms and helped generate the enormous diversity of plant life on Earth. Conversely, it follows that had large-scale genome duplication events not occurred, the world would be a very different place. The kingdom of plants, as we know it, would probably not exist, and neither would vertebrates. Aided in part by whole genome duplications at critical moments in the past, plants survived the repeated mass extinctions that befell the animal world over the past half billion years. Quite how plant life managed the impressive evolutionary trick of co-opting and redeploying an expanding genetic repertoire for their own ends to build leaves and roots out of bits of DNA inherited from algae is a story for Chapter Four.

# ANCIENT GENES, NEW PLANTS

'The attempted synthesis of paleontology and genetics.... may be particularly surprising and possibly hazardous.'

George Gaylord Simpson, *The Tempo and Mode of Evolution*, 1944

Botanic gardens are often steeped in history, places where plants and the past combine to enrich the magic of a visit. Founded in 1621, the Botanic Garden of the University of Oxford is the oldest in Britain, a mere 76 years younger than the oldest botanic garden in the world: the Garden of Padua in northern Italy. Founded in 1545, the Garden of Padua originated from a Benedictine collection of medicinal herbs. Now a UNESCO world heritage site, it retains the original layout with a circular central plot, symbolizing the Earth, surrounded by a moat. Similarly, the enchanting Oxford Botanic Garden also preserves its original layout. Pass through a classical baroque-style stone gateway and you find the walled garden that originated in the seventeenth century, arranged around the oldest tree, a magnificent creaking yew (*Taxus baccata*). The tower of Magdalen College looms over the gardens, and each year the college choir celebrates the arrival of spring at 6am on May Day morning, when their voices rise high and clear singing a Latin hymn.

For many years, an iconic black pine (*Pinus nigra*) stood towards the far side of the walled garden. With its thick twisted branches snaking upwards carrying a bristling canopy of needles, the tree was a favourite of the author J.R.R. Tolkien (1892–1973) and may have inspired his epic *Lord of the Rings* trilogy. One of the last photographs taken of Tolkien has him sitting companionably against its trunk.[1] Sadly, though, in 2014 the decision was taken to cut down 'Tolkien's tree' after two of its limbs creaked, groaned, and then finally collapsed, making it unsafe for visitors to the garden. Hope is not lost, however. Future generations, and Tolkien fans, may yet get the opportunity to see it, as there are plans to attempt propagation of the magnificent tree from cuttings.

All told there are around 3000 botanical gardens worldwide, living repositories of plant diversity that provide documented collections of plants for our enjoyment, as well as for conservation and research.[2] Over half of the world's population now lives in urban areas and botanical gardens provide them with green oases (**Figure 8**). The telling statistic is that 250 million people visit these botanic gardens each year. It points to our deep-seated desire to experience the greenery missing from our sophisticated urban lives and, perhaps, kindle an interest in understanding something of its evolutionary history. Clues to the past are there for those who care to take a close look at the plants themselves. Notice the underlying unity in the way they are constructed. Everything, from ferns and trees to shrubs and grasses, conforms to the same basic modular body plan of repeated structures and organs that form shoots and roots. In fact, when you stop and think about it, it is easy to appreciate that shoots are simply composed of bits of stem and leaf, iterated repeatedly until a flower or some such reproductive structure appears on the end. Land plants hit on this winning blueprint early in their evolutionary history and have stuck with it ever since.

Plants began colonizing the land back in the Ordovician (named after a Celtic tribe living in Wales, the Ordovices), and their explosive diversification across the landscape followed in the Devonian Period (419–355 million years ago). Written in rocks of Devonian age is a geological diary beautifully documenting the process of green hegemony that ensured the irreducible dependency of the world on the photosynthetic products of plants. The habitats and ecological opportunities created by evolving terrestrial floras set the stage for the subsequent appearance and diversification of animal life on the continents. Irresistibly dubbed the Devonian 'big bang', this remarkable burst of evolutionary innovation saw the escalating ascendency of plant life over an interval of roughly 60 million years.[3] The Devonian Period was named after the picturesque county of Devon in south-west England by geologists who first studied the fossil-bearing deposits in its rocks. We can liken the plant kingdom's Devonian 'big bang'[4] on the land to the earlier 'Cambrian explosion' of the animal world in the oceans. The Cambrian explosion began about 541 million years ago and saw most of the great animal phyla appear in the fossil record over about 50 million years—suddenly, in geological terms. Both events have the hallmarks of dramatic biological innovation at the start, and

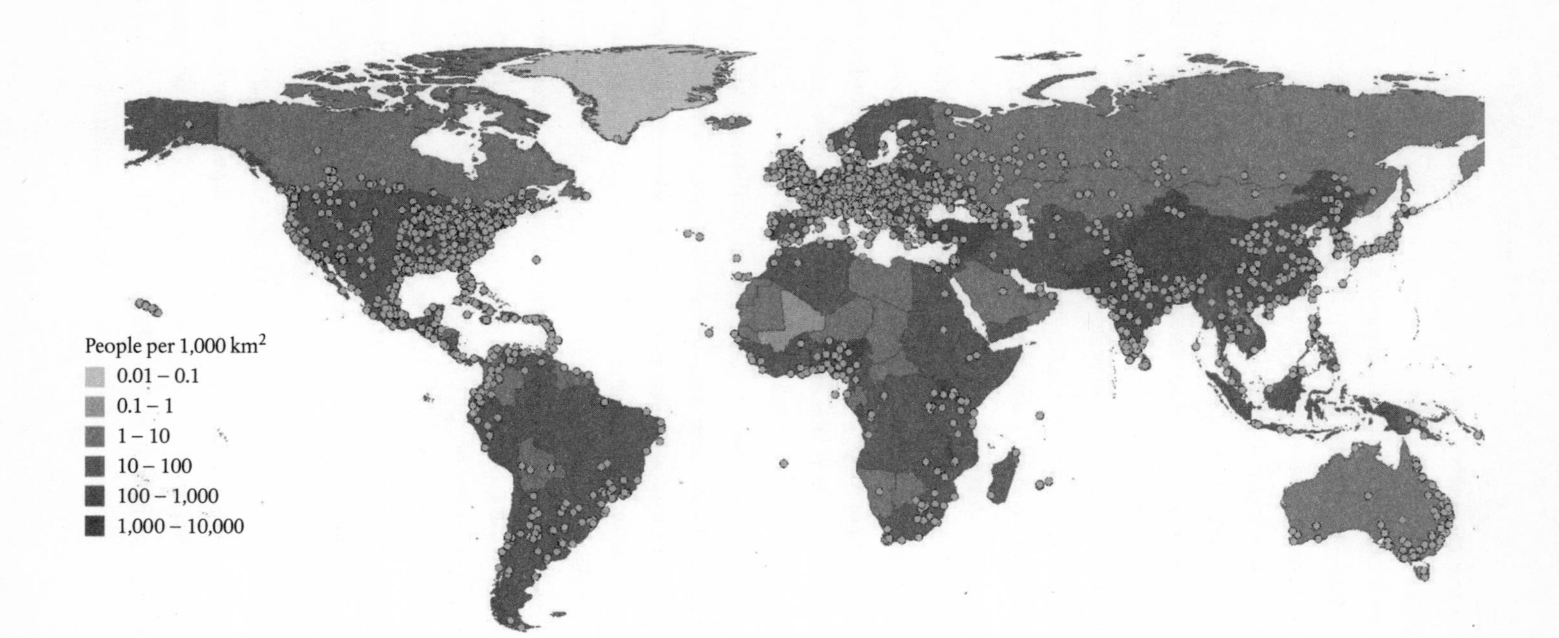


**Figure 8** Global network of over 3000 botanic gardens overlain onto a map of human population density.

both ended with the assembly of a complex web of life, forming ecosystems broadly analogous to their modern counterparts.[5] By the end of the Devonian 'big bang', most of the major modern lineages of plants—bryophytes, lycophytes, ferns, and seed plants—had originated, and with them striking organs and tissues like leaves, roots, seeds, and wood.

Palaeontologists love to study the fossil record of this time, and for good reason. A handful of sites continue to offer rich and rewarding new sources of information on the evolution of the plants themselves. A few exceptional localities have fossils in which soft tissue has been preserved, a consequence of unique conditions under which minerals have replaced the original organic matter to record astounding details of cellular anatomy. Still, the fact is that neither palaeontologists, nor anybody else for that matter, can do experiments with the fossils to discover the details of how plants underwent such dramatic diversification in form and function. Fossils provide the framework for thinking about what happened, but stay silent on the matter of *how* it happened. For insights into how the drama unfolded in the evolutionary theatre of the Devonian world, we need to turn to the DNA record of living plants.

We have already seen how DNA helped solve the question of the origin of land plants: they evolved from freshwater charophyte algae that surmounted the formidable challenges of life out of water. We have also seen how the genomes of wayfaring plants changed during their spectacular transition from anatomically simple freshwater algae to large, complex, multicellular organisms such as trees, cloaking the near-barren landscape with forests of leaves. Now we must concern ourselves with how the genetic 'toolkits' within those plant genomes built the innovations they needed for succeeding in the new rocky environment in which they found themselves. We must give substance to claims that ancient genetic programmes became co-opted, expanded, and were repeatedly fine-tuned by evolutionary rewiring to give rise to the amazing diversity of plants alive today.[6] This is really the everyday stuff of evolution revealed by a branch of science called 'evo-devo'. Evo-devo deals with the evolution of developmental pathways regulating how plants (and animals) grow cell by cell from embryo to adult. By seeking to understand how the genetic circuitry of the cells operates, it provides fundamental insights into the nature of evolution itself. It seeks to document the Darwinian processes of natural selection and descent with modification across vast tracts of geological time.

One of the first things the fossil record makes clear is a surprise: shoots evolved before roots. Leafless shoots came first, followed by leafy shoots, followed by shoots with proper roots. The term 'leafless shoots' might be over-cooking it as a description of the earliest vascular land plants. These tiny plants that lived on land over 400 million years ago were little more than simple naked delicate axes, only a few millimetres tall, anchored into sediments with simple fine rootlets. Presumably, the naked stems of these minute photosynthetic pioneers were green, perhaps like those of modern whisk ferns (*Psilotum*) or sedges (*Carex*), to allow them to conduct the daily business of photosynthesis, although there is no way currently of knowing for sure.[7] Whatever the extent of 'green-ness', these land plants photosynthesized without leaves. That innovation appeared in its simplest form as a microphyll leaf, essentially a narrow strap-like structure with a single unbranched vascular strand running through the centre, which appeared a few millions years later. Lycophytes had microphyll leaves and later gave rise to the giant trees that grew to heights of over 40 metres and dominated the Carboniferous swamp forests 300 million years ago. Today, this type of leaf is only to be found clothing the miniature cone-bearing spires of the last surviving lycophyte lineages from those early days of plant life on land: clubmosses, spikemosses, and quillworts. Quillworts are small aquatic herbaceous plants that also possess another overlooked trait connecting them to the giant trees of the Carboniferous—their rootlets share a patterned architecture similar to those found in fossils of the long-extinct trees.[8]

Leaves of a sort that are familiar to us, those with an arched flat blade and a cantilevered support structure, begin to turn up in the fossil record 30 million years after plants with microphyll leaves appeared on the scene. These laminated structures are called megaphyll leaves. A megaphyll leaf is really a photosynthetic leaf blade irrigated with a complex, branching vascular network of tubes for efficiently delivering water to the photosynthesizing tissues.[9] Among the oldest examples are those of the rare specimen called *Eophyllophyton bellum*, whose fossils have been excavated from 390-million-year-old Devonian rocks in China.[10] The exquisite, deeply incised miniature leaves of *Eophyllophyton* fossils are preserved as thin films of carbon, black smears on rock surfaces reaching only a few millimetres across. The quality of preservation of these early Chinese fossil leaves is astonishing. Prepare them with sufficient care and patience, and it is possible to reveal intricate networks of branching veins within the fossilized tissues. By late in the

Devonian, everything from ferns to seed plants converged on leaves as an optimal solution for harvesting solar energy to power photosynthesis.[11] The fossil record tells us that at least four different vascular plant lineages independently evolved the same variety of solar umbrella, including progymnosperms (extinct forerunners of gymnosperms), sphenopsids (extinct), ferns, and seed plants. Each of the four lineages then followed the same evolutionary trajectory for making complicated leaf shapes, a fact that genetics can explain, because all four inherited the same genetic toolkit for programming leaf development from a common ancestor, and there were only so many ways it could be rewired to make complex leaf architectures.[12] Leaves, then, have a certain inevitability about them. Re-run the evolutionary tape of life, to use a metaphor beloved of some evolutionary biologists, and it is a safe bet that sooner or later photosynthetic life forms colonizing the land will start making leaves.

Yet the fundamental question is surely this: what was the underlying genetic machinery that gave rise to leaves in the first place? How did plants evolve from tiny naked stems into leafy shoots? Answering that question means that we first need to know something of how modern plants make leaves, and the action starts with stem cells located at the apex of the shoot. Stem cells are special and, rather confusingly, do not simply make stems. They provide a source of new cells to grow or replace a variety of specialized tissues. To perform this function, they divide to renew themselves, leaving their descendants free to accumulate and form the new tissues. It follows, then, that stem cells possess the special property of being able to produce the full diversity of cell types a plant needs for growth and reproduction. Indeed, this was confirmed by classic experiments in the 1970s which showed that virtually all the tissues of a shoot descend from small groups of stem cells. This explains why entire plants with stems, branches, leaves, flowers, and so on can be grown from small pieces of tissue, or even a single cell.[13] It also explains why all is not lost for Tolkien's tree; regeneration ought to be possible. Stem cells are really the secret to the long lifetimes of trees. They can regenerate and produce populations of unspecialized stem cells over the lifetime of the plant to renew worn-out structures, shielded, somehow, from mutational meltdown. Anyway, the point is that leaves arise from the cluster of self-renewing stem cells at the tip of the shoot, called a meristem (**Figure 9**).

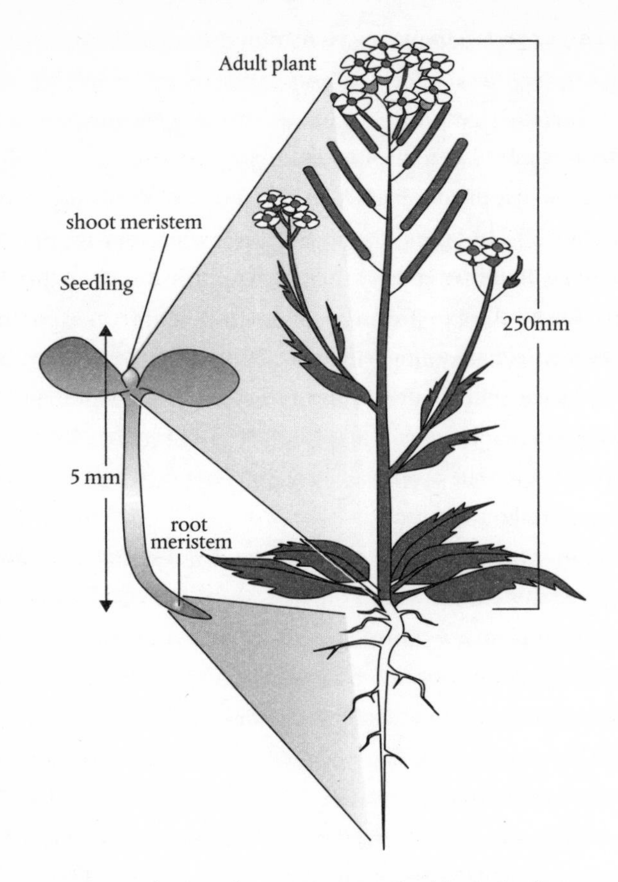


**Figure 9** Shoot and root meristems build whole plants.

One of the tasks of meristems in the shoot is to constantly 'decide' when it is appropriate to continue growing and when to do some other more specialized job like building a leaf or flower. Genes tightly controlling the balance between normal shoot growth and making leaves have a shared ancestry that reaches back to the golden age of Devonian evolutionary innovation.[14] Tight regulation is essential. If just one extra 'accidental' cell division occurred per year, shoot meristems could be a thousand times larger than normal within a decade,[15] and trees that live for many decades would become grotesque super-organisms. A commonplace example illustrating the urgent need to regulate when to grow a leaf, and when to do some other more specialized job like building a flower, is the case of cauliflowers. The head of a cauliflower shows what happens when meristems go wrong, as masses of flower buds fail to undergo normal floral development, causing over-proliferation of cells and the formation of the thick creamy 'curd'. It is possible to induce a similar mutation in the flowering plant *Arabidopsis* and make it produce weird miniature 'cauliflower-like' structures on its flowering shoot.[16] The signals that determine when the meristem 'decides' it is time to produce daughter cells destined to become leaves are not fully understood, but we know they involve the powerful plant growth hormone auxin (from the Greek *auxein*, meaning to grow).[17] Whatever the mechanism, leaf formation is initiated at specific positions in the shoot tip and at regular intervals, as revealed by an often helical sequence of bulges that form around the meristem. This ensures the appearance of the elegant and efficient arrangement of leaves that minimizes competition for light between them.

Leaf formation is initiated when the meristem transitions from indeterminate growth to a determinate growth programme. In indeterminate growth mode, plants can become giant trees like those of the primeval swamp forests, but in determinate growth mode things become finite. Determinate growth is the norm for all animals, including humans. Our bodies call it a day and stop growing at maturity and in this sense we are just like plants when they make leaves. This transition requires switching off genes maintaining self-renewing stem cells in the shoot meristem, and activating those directing the development of leaves. The *KNOX* family of genes maintains cell production in the shoot apex by coding for a suite of powerful transcription factors. These are small proteins that bind to each other and specific segments of the DNA molecule to switch genes on and off at specific times and places. In this way, transcription factors provide plants with a set of logical molecular operators for regulating patterns of gene expression that control their development. If factor X binds to factor Y, then the resulting combination may bind to the DNA molecule in a particular region and switch a single gene, or whole collections of genes, on or off. The default setting for *KNOX* genes is 'activated', meaning they normally produce transcription factors blocking the development of new leaves. During leaf development, proteins coded by another set of genes—*ARP* genes—block the transcription of *KNOX* genes, allowing recruitment of stem cells into the developmental pathway for making leaves. Because of this negative interaction, *ARP* gene proteins are said to 'repress' the activity of *KNOX* genes. *KNOX* genes derive their name from the knots they produce in leaves when plants are engineered to express them in leaves instead of the normal shoot meristem position. Knots appear because cells undergoing excessive divisions cause the rapidly expanding leaf blade to pucker and buckle.[18]

Originally discovered in the crop plant maize,[19] the *KNOX–ARP* genetic toolkit for instructing shoots to make leaves was discovered later to be doing much the same role in the flowering plant *Arabidopsis*. This tells us something important. Because according to the green tree of life, these two species last shared a common ancestor over 140 million years ago. The logical inference we can draw from this is that the *KNOX–ARP* two-component core genetic toolkit for making leaves was inherited from a common ancestor that lived at that time, or even earlier. The question is how much further back in time can we trace the evolutionary history of this genetic toolkit for building leaves?

Jane Langdale is a British professor of molecular biology addressing exactly this sort of question by scrutinizing the evolutionary history and function of genes involved for everything from mosses and hornworts to ferns and flowering plants. Langdale is unusual for starting out her career as a human geneticist before being won over by the charms of the plant kingdom. She could not resist being seduced by big science questions about the origin and diversification of our land floras, and about the rich opportunities for seeing further by marrying evidence from fossils and DNA molecules. Working with the spikemoss *Selaginella*, her team reported the exciting discovery a decade ago that the *KNOX–ARP* genetic toolkit for building leaves had already originated by some 350 million years ago, far earlier in plant evolutionary history than anyone had previously suspected. The evidence came from elegant gene-swap experiments. At the time these were regarded as cutting edge, but they have since become routine and the cornerstone of unpicking the antiquity of genetic programmes governing plant development.[20]

The trick leading to this discovery was to engineer lines of *Arabidopsis* in which the *ARP* gene has been deleted from the genome, called knock-out lines in the language of geneticists. These lines conspicuously failed to produce normal leaves arranged in compact rosettes around the base of the plants. Unable to repress *KNOX* activity and allow stem cells in the shoot meristem to transition towards leaf formation, these lines developed strangely deformed leaves with thick stems and lacking proper leaf-blades.[21] However, substitute the *ARP* gene from *Selaginella* into these *Arabidopsis* lines and more-or-less normal leaves developed.[22]

In essence, these experiments demonstrated that the *ARP* gene important for making microphyll leaves in a lycophyte functions interchangeably with its counterpart in a flowering plant for making megaphyll leaves. The magic of doing science is felt keenly with such breakthrough moments that open the door to understanding the origin of leaves themselves.

In short, then, what we have seen so far is that two lineages of land plants, separated by more than 350 million years of evolution, possess common genetic toolkits for building different types of leaves (microphylls and megaphylls). In fact, the discovery has greater evolutionary significance because the common ancestor of lycophytes and flowering plants lived before leaves originated. Our best guess suggests this was an ancestral leaf-less twig-like plant. If correct, then the *KNOX–ARP* module inherited by both lineages was part of the core genetic toolkit of the earliest leafless vascular land plants. This offers us an insight into how evolution works. Lycophytes and flowering plants might have evolved different mechanisms for repressing *KNOX* activity to make leaves. Instead, evolution co-opted a pre-existing genetic module, which originally regulated how the naked shoots of simple early land plants grew, and redeployed it to control the developmental programme for making leaves.

We should not be surprised to learn that the evolution of leaves involved more genetic components than the ancient *KNOX–ARP* module, and new discoveries are revealing once again the deep roots of that machinery. Leaves are programmed to develop in size and shape in three dimensions—the width, thickness, and length of a leaf blade are all genetically predetermined. Then there is also the essential requirement for the developing leaf to synthesize the correct sorts of cells and tissues in the right place and at the right time. The surface of a leaf facing upwards towards the sky, for example, is specially built for intercepting sunlight. In consequence, it has densely packed cells filled with chloroplasts for capturing light energy. The lower leaf surface is built with stomata, the microscopic gas valves that facilitate capture of carbon dioxide from the atmosphere for photosynthesis. Leaves also require the synthesis of vascular tube networks as the structure develops for transporting water and sugars.

Many of these complex aspects of leaf development are under the control of an ancient family of genes (snappily named *HD-ZIP* genes) that originated in green algae long before land plants evolved.[23] Based on their exceptional antiquity and great utility when it comes to governing plant development, *HD-ZIP* genes were probably part of the basic genetic toolkit of early land plants. In these plants, they were probably originally involved in orchestrating the formation of vascular tissue in stems. In modern lycophytes, for example, *HD-ZIP* genes regulate the development of vascular plumbing in their stems and microphyll leaves. This suggests that lycophytes gained the distinguishing vascular strand running through their leaves by recruiting similar genetic machinery that was already responsible for producing vascular tissues in stems. The great plant morphologist F.O. Bower would have been pleased with this. Some 80 years ago, he proposed that microphyll leaves arose through modification of spiny outgrowths protecting plants from attack by herbivorous animals.[24] Now the genes have spoken, it seems he may have had a point. Only hundreds of millions of years later did *HD-ZIP* genes acquire new roles in specifying the identity of cells and tissues forming specialized megaphyll leaf surfaces.[25] The story of *HD-ZIP* genes illustrates how, under the right selection pressures, pre-existing components of a charophyte algal genetic toolkit can be redeployed for making novel structures that algae lack—leaves.

Complementing our emerging picture of the genetic machinery for building leafy shoots are discoveries from the world of molecular genetics that explain how roots evolved. As with leaf evolution, the fossil record establishes the framework for our thinking.[26] When early terrestrial floras colonized Earth's continental surfaces, fossils suggest they had nothing in the way of proper roots. Instead, the first land plant rooting structures were basic filamentous cells called rhizoids. Rhizoids probably originated over 470 million years ago in freshwater algae,[27] with the oldest example of a fossil land plant having simple, shallow, rhizoid-based rooting systems dating to the early Devonian, 411 million years ago. By analogy with the function of rhizoids in living plants, it seems reasonable to assume they performed a similar job in these early land-based photosynthetic pioneers, nourishing shoots by delivering flows of water and dissolved nutrients from sediments and soils.[28]

By the time short herbaceous vegetation appeared later in the Devonian, spidery rhizomatous systems had developed that helped stabilize river banks and create soils by providing an organic matrix for binding everything together and resisting erosion.[29] Some of the best examples of these simple structures captured in the rock record belong to an extinct fossil plant called *Asteroxylon* (**Figure 10**).[30] The appearance of the parts of *Asteroxylon* that were below ground is markedly at odds with that of its leafy shoots. This distinction immediately tells us that

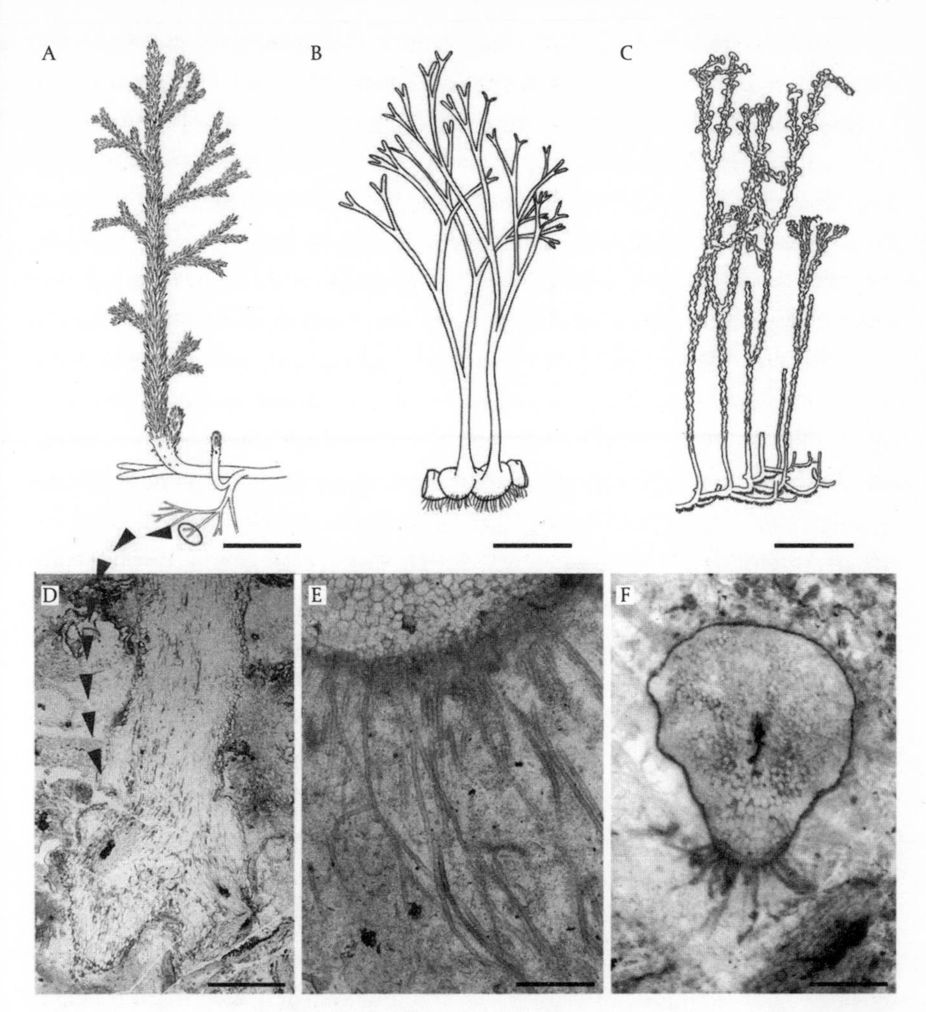


**Figure 10** Early Devonian land plants and their rhizoid rooting systems. (A) *Asteroxylon mackiei*, (B) *Horneophyton lignieri*, (C) *Nothia aphylla*. D to F, Rooting structures. (D) Longitudinal section of the rooting system of *A. mackiei* showing dichotomized branching. (E) Transverse section of a corm bearing rhizoids of *H. lignieri*. (F) Transverse section of a rhizome of *N. aphylla* showing a ridge on the ventral surface that bears the rhizoids. Bar = 4 cm in (A), 3 cm in (B) and (C), 1 mm in (D), 0.45 mm in (E), and 1.5 mm in (F.)

Devonian plants possessed diverse genetic toolkits for generating specialized tissues composed of organized communities of cells needed for their very different above- and below-ground parts.

In the terms of the plant development scheme described earlier for shoots, we know that roots arise from another cluster of stem cells located at the root tip. The root meristem has a somewhat simpler task than its above-ground counterpart does, in that it only has to give rise to the tissues of the main root. In other words, a wonderfully complex plant can develop all of the intricate tissues it needs from a handful of stem-cell clusters in shoots and roots (**Figure 9**). Several different groups of vascular plants independently evolved rooting systems, including seed plants, ferns, horsetails, and their allies, and they took on a great diversity of forms. Remarkably, a tiny fossil root tip, just 2 mm across, belonging to a giant lycophyte tree, has been discovered in Carboniferous rocks (ca. 320 million years old).[31] This extraordinary fossil preserves beautiful details of cells in an actively growing root meristem and shows us that this ancient root tip is organized much like that of living gymnosperm roots. With the later appearance of trees like *Archaeopteris*, increasingly deep and elaborate rooting systems took hold through the Devonian as diverse forest ecosystems tethered themselves to the continents. The evolution of large and complex root structures adorned with fine root hairs must have provided these trees with improved mechanical anchorage into soils as well as a greater capacity for extracting water and scavenging nutrients required to support their prodigious growth.

Thanks to insights from these sorts of fossils, the task of reconstructing the genetic instructions early land plants might have used to make their simple rooting systems now becomes a little more straightforward. It amounts to unpicking the genetic pathways controlling the development of rhizoids in mosses and liverworts, and comparing them with those that flowering plants deploy to make hairy roots. Oxford University-based geneticist Liam Dolan has dedicated his scientific career to this worthy quest. He is concerned, some might say obsessed, with all things rooty—from genes to fossils. He provides a sort of complementary below-ground counterpart to Langdale's work on shoot evolution at Oxford.[32] Dolan has discovered that the pair of genes regulating the formation of multicellular rhizoids in moss is the same as those making the fine single-celled root hairs adorning the roots of the flowering plant *Arabidopsis*.[33] The story leading to this discovery starts with what we know about where root hairs in *Arabidopsis* roots originate. The cells on the surface of the root are organized in alternate rows of hair-forming cells and non-hair cells. Each root hair is a single cell, which produces a tip-growing tubular protuberance that becomes a hair. Two

*RSL* genes control the process. The two genes are very similar to each other in terms of the amino acids they code for, suggesting that one is a copy of the other that arose by a gene duplication event. Activation of these genes produces proteins that accumulate in the hair cell and trigger additional genes to grow a root hair. Delete either gene from *Arabidopsis* and the resulting plants develop naked, hairless roots. Mosses grow filamentous multicellular rhizoids instead of root hairs under the guidance of a pair of genes related to those used by *Arabidopsis*, and if you delete them in knock-out lines, the moss develops stunted rhizoids.[34]

In the next step, Dolan's group transferred one of the moss genes into *Arabidopsis* lines lacking their own copy of the equivalent gene; they then gain the ability to develop normal, hairy roots. Note that 'gene-swap' experiments from an ancient to a more recently evolved plant lineage is the same strategy as that which Langdale adopted in her work on the evolution of leaves, but with a different group of plants. Langdale used a lycophyte, and Dolan used a representative of the non-vascular bryophyte group, mosses.[35] What these experiments demonstrate is that the genetic machinery mosses use for building multicellular rhizoids can also control the development of single-celled root hairs in flowering plants. Given these two lineages last shared a common ancestor over 400 million years ago, the implication is that these genes were present in the earliest land plants and enabled them to grow rhizoids.

As is so often the case in science, when you delve into the details, things turn out to be more complicated than they first appeared, and there is actually a network of genes acting at different stages of root hair and rhizoid formation. Genes in the moss rhizoid network are able to act interchangeably with those of the flowering plant root hair network,[36] but there are fundamental differences in how the genes are wired together which allow the growth hormone auxin to work in different ways.[37] In moss, auxin promotes rhizoid development by acting on individual genes independently of the others in the network. In a flowering plant, however, it acts on one set of genes in the network, which in turn regulate the activity of another set.[38] What we have, then, are similar genes controlling the development of root hairs and rhizoids that are wired up differently to produce different structures. Intriguingly, auxin also promotes the growth of rhizoids in charophyte algae.

## THE 'HAND OF HISTORY' FAILS TO HALT 'THE TIDE OF PROGRESS'

Liam Dolan holds the Sherardian Chair of Botany at Oxford University, once occupied by the plant geneticist Cyril Darlington (1903–1981). Darlington discovered chromosomes, but his excursions into eugenics involved him in numerous disputes and isolated him from mainstream science.[39] Gifted and controversial, Darlington was tall and handsome with film star looks in his younger days—think Errol Flynn and you get the idea. Darlington remorselessly disparaged his colleagues' work if they dared to engage in any branch of botany other than his beloved genetics. He sniped at his distinguished ecological colleagues by paraphrasing Oscar Wilde with the scathing quip that ecology is the 'pursuit of the incomprehensible by the incompetent'.[40] Darlington was fascinated with the Botanic Garden at Oxford and introduced genetic hybrids important for his chromosome studies. Search the grounds and you might find the date-plum tree (*Diospyros lotus*) planted to commemorate his memory. In the twenty-first century, under pressure for space as facilities expand, Darlington's wood-panelled office became swallowed up by modernity.

Office pedigree in academia is undoubtedly a niche subject and we should not become distracted by it, except to note that a similar story unfolded for the office facing mine in Sheffield. The lineage of its distinguished former occupants reaches back to at least 1938 with the leading plant ecologist Roy Clapham,[41] our erstwhile head of the Department of Botany (as was) in Sheffield for 25 years, who had arrived from Plant Sciences in Oxford. David Walker, who laboured on elucidating important biochemical details of photosynthesis, followed Clapham.[42] Walker was an eccentric. He had an extra door built into his office wall to permit rapid escape into his adjacent lab should earnest students come looking for their tutor. The distinguished mycologist Sir David Read, who went on to become Vice President of the Royal Society, followed Walker. Walker's escape hatch has since been covered over during its conversion into office space. Everywhere, it seems, the remnants of scientists past are disappearing.

This is not the end of the line for Dolan's *RSL* genes, however, because they also control the formation of specialized cells and structures in liverworts.[43] As we have already seen, fossils suggest that liverworts are living representatives of an ancient plant lineage. They may have ancestral attributes representative of the early land plants colonizing terrestrial environments,[44] although such inferences are contingent on resolving uncertainties in the evolutionary relationships between the bryophytes (hornworts, liverworts, and mosses) and vascular plants.[45] Nevertheless, the point is that the liverwort *Marchantia polymorpha* anchors itself into the ground with single-celled filamentous rhizoids whose development is under the control of *RSL* genes. These genes also function to make asexual multicellular reproductive propagules (called gemmae) held in splash cups on the upper surface of the plant, and multicellular slime cells, called papillae. *Marchantia* lines lacking *RSL* genes fail to develop rhizoids, have empty gemmae cups, and lack papillae (**Figure 11**). Conversely, engineer plants with a hyperactive *RSL* gene and the liverwort sprouts hairy rhizoids on both its upper and lower surfaces, and a super-abundance of gemmae. So it seems that *RSL* genes can control both the formation of unicellular structures, such as rhizoids for nutrient acquisition, and multicellular structures for reproduction. Liverwort *RSL* genes also function perfectly well when transferred to flowering plants lacking *RSL* genes, enabling them to grow hairy roots.

The startling conclusion drawn from these discoveries is that *RSL* genes likely controlled the development of the first land-plant rooting systems. Indeed, they probably already existed in the charophyte algal ancestors of land plants, which, for good measure, also passed on mechanisms for taking up the essential inorganic nutrient phosphorus from soils and sediments.[46] Certainly, the advantages hairy roots give to plant life when it comes to succeeding on land are not to be underestimated. Rye plants typically have around 14 billion root hairs, and these provide a surface area of over four hundred square metres for nutrient and water uptake.[47] Multiply such benefits up as land floras sank their rooting systems into developing soils, and it is easy to imagine how root evolution must have played a crucial role in shaping the green future of Earth's continents. We can go further than this, because the transition from freshwater to a terrestrial existence involved the evolution of a number of specialized cells, tissues, and organs required for survival and reproduction, and these are ancestral functions of *RSL* genes.

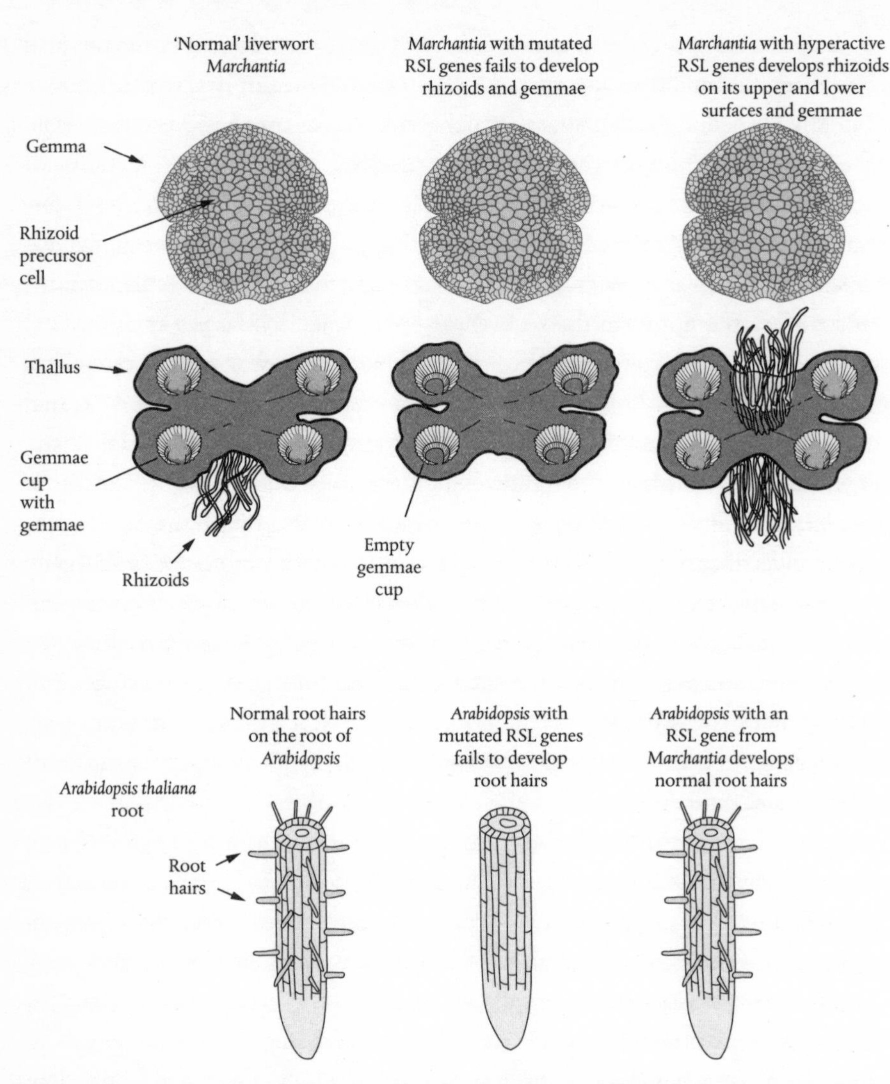


**Figure 11** Ancestral functions of RSL genes in land plants as demonstrated by genetic manipulations with a liverwort (*Marchantia*) and a flowering plant (*Arabidopsis*).

Dolan's group are keen to exploit these discoveries by showing how fundamental plant science research can make a wider contribution to society by tackling issues related to food security in a changing climate. Engineering crop plants with hairier roots means they might be better able to withstand the heat and drought of warmer future climates by taking up more water from the soil. The genomes of grasses such as rice, wheat, and maize also contain *RSL* genes that control root hair formation and function interchangeably with flowering plants.[48] This implies that their function has remained similar (or conserved) for millions of years since the evolutionary split between monocots and dicots. Eventually it may be possible to manipulate the *RSL* genes in wheat and rice to develop super-hairy roots.[49] Such genetically 'improved' plants might grow bigger and produce higher yields, yet require fewer artificially provided nutrients because they are more effective at scavenging them from the soil.[50] Similar arguments hold for water, bringing within reach improved agricultural food production with reduced demands for artificial fertilizers.

We are starting to appreciate how evo-devo studies into the genetic workings of living plants shines a light on the genetic make-up of ancestral plants that have been extinct for hundreds of millions of years. Fossils in rocks archive the dramatic diversification of plant life, and DNA connects us to that vanished world by explaining how the molecular events inside cells brought it about. There is one further important point uniting evidence from fossils and molecules that we need to unpack, because it helps explain a little more how tissues in the sporophyte life-cycle phase may have arisen.

Recall that extant plants have two different life-history phases. For example, the moss gametophyte (green filaments) and sporophyte (spore capsules) or, in the case of ferns, the diminutive gametophyte (called a prothallus) and the large sporophyte with its arching fronds. Now, fossils suggest that, unlike these living examples, the tissues of both life-cycle phases in early vascular land plants may have been complex. To be sure, the two phases were not identical and the habit and size are poorly understood, but it seems the gametophyte and sporophyte bore greater similarity to each other than is the case for living species. They also developed a greater diversity of tissue types, including rooting structures, a vascular system, and microscopic pores called stomata. And this has far-reaching consequences for understanding plant evolution because it implies that hundreds of millions of years ago the genetically controlled developmental programmes for making gametophytes and sporophytes were similar.[51] The evolution of an independent sporophyte, which dominates the world's floras today, was made possible by the co-option and redeployment of an existing repertoire of developmental programmes that the gametophyte was already using to live as a fully independent plant. This hypothesis infers genetics from fossils and is supported by the findings from evo-devo we have been considering. Many ancient genes and gene regulatory networks discussed—from leaves, shoots, roots, and vascular tissues—were recruited from pre-existing networks that operated in the gametophyte generation.

Still, comparatively speaking, important innovations like leaves and roots are simple structures. So what does evo-devo have to say about how plants pull off the phenomenal trick of developing two completely different generations within a single lifetime? Botanists call this kind of split life cycle the 'alternation of generations', and it means that a single genome has to code for two developmental programmes: one for building the sexual generation (a gametophyte plant) and one for building the asexual generation (a sporophyte plant). Until recently, geneticists and developmental biologists working on the evolution of life cycles were stumped when it came to explaining how plants managed to achieve this remarkable feat. Then a recent series of breakthrough discoveries, in a variety of organisms ranging from fungi and algae to moss and flowering plants, transformed our understanding by showing that they all share common genetic machinery controlling the transition from one developmental stage to the next.[52]

Our understanding of land-plant life cycles comes mainly from research on moss, and it is a story about gene silencing to completely repress the gametophyte developmental programme in order to make a sporophyte and vice versa.[53] The gene responsible for controlling the development of sporophytes in moss is called *MKN6*. It works by suppressing the developmental programme for making the green filamentous strands, i.e., the gametophyte generation.[54] Put another way, *MKN6* is a gene silencer, blocking gametophyte development to enable the sporophyte to develop instead. It is a similar story in ferns, although the details are not nearly as well understood. Ferns are the evolutionary intermediate lineage between mosses and seed plants. The fern *Ceratopteris* has four genes closely related to *MKN6*, activated only when the adult (sporophyte) generation is developing.[55] Together these observations suggest that the fern genes switch off the developmental programme of the gametophyte. Mosses and ferns seem to have adopted similar gene-silencing mechanisms for releasing the developmental programme to make the sporophyte stage of their life cycles. The switch is of considerable antiquity, dating back to the time of the last common ancestor of eukaryotes.[56]

Now consider the opposite situation, the formation of the gametophyte generation. This requires inactivation of the sporophyte developmental programme, this time through a set of mechanisms involving epigenetic gene silencing.[57] Epigenetics is a fashionable branch of science that does not involve modification to the DNA coding for genes but instead changes to how genes are expressed, or indeed, if they are expressed at all. In the real world, gene expression is not really an 'on' or 'off' toggle switch; the levels of gene expression vary, like the analogue volume control of a radio. Epigenetic modifications allow plants to vary the volume control of genes and can involve chemical modification to the globular proteins around which DNA is wrapped inside the nucleus of a cell. It is actually rather misleading to think of the DNA double helix as snaking around inside a nucleus. In reality, it is wrapped tightly around protein globules that are stacked on top of each other, condensing the volume of space the DNA molecule occupies. When a cell divides, chemical modification to these globular proteins, called histones, can be passed on, so control of gene expression stays consistent from mother cell to daughter cell.[58]

The details are complex but the bottom line is that two genes are involved in epigenetic silencing of the developmental programme responsible for building moss sporophytes (for the record, they are called *FIE*[59] and *CLF*[60]). These genes code for a protein that catalyses the chemical modification of a particular target histone, which blocks access to the DNA at key points. Mutations in either gene cause the filamentous gametophyte to develop spontaneously sporophyte-like bodies without the usual need for fertilization of the egg cell by sperm. *Physcomitrella* mutants lacking *CLF*, for example, have gametophytes relieved of epigenetic 'silencing', which then develop sporophyte-like structures.[61] Under laboratory conditions, these mutants can exhibit branching patterns reminiscent of *Cooksonia* fossils.[62] Components of the epigenetic machinery may have originated in algal ancestors, which then increased in complexity as plant life transitioned from water to land sometime in the Ordovician.

We are starting to see how ancient genetic machinery enabled the evolution of leaves, roots, and complex life cycles that lay behind the great Devonian diversification in the kingdom of plants.[63] But as plants with increasingly elaborate architectures sprang up and began photosynthesizing in the sunlight, all this potential for riotous growth meant they needed the ability to start making reliable decisions about when to grow, and in what direction. Chemical messengers bringing news from the outside world, in the form of hormones, are the answer. Synthesized in response to environmental cues, and transported between different parts of the plant, hormones tell shoots or roots when and where to grow. Foremost among these hormones is auxin. We have already touched on its role in leaf and root development, but it actually regulates almost everything in plants. Writing *The Power of Movement in Plants* in 1880, Charles and Francis Darwin postulated its existence nearly 50 years before the Dutch botanist Frits Went (1903–1990)[64] named and chemically characterized auxin in 1926. Auxin is the mysterious 'power' of plants in the title of their book.

The role of auxin biology in orchestrating the evolving complexity of the vegetable world's takeover of the land is far from understood. Most green lineages have genes for synthesizing it,[65] but what is important for seeing further is knowing about the shared heritage of the genes involved in transporting auxin. Modern plants produce auxin in their shoot tips and transport it downwards from the shoot to the root via files of specialized cells in the vascular tissues. This establishes a directional flow of auxin and creates concentration gradients within the different parts of the plant, which are the key to its widespread effects. Embedded in the walls of the cells transporting auxin are special membrane-spanning proteins that regulate the flux of auxin into and out of cells. Genes make the proteins and we can trace their evolutionary history to see when they originated. Traced in this way, we find that genes coding for the proteins involved in transporting auxin are ancient, originating sometime before green plants made the transition to land.[66] Exactly how they arose is unclear, but one idea is that the ancestral function of the transporter proteins was to regulate cellular metabolism by sequestering auxin inside cells.[67] To do this, the proteins were highly localized within the membranes of special compartments located inside the cells of the immediate algal ancestors of land plants. At some point in time, an important mutation then changed the localization of these proteins, moving them to the cell wall. With that development, plants gained the ability to export auxin from cells and establish directional auxin transport, a necessary step for achieving its action-at-a-distance properties.[68]

Directional auxin transport can be seen in trees of the early Carboniferous forests, which have not only been preserved as fossil wood, but also carved in stone at the Natural History Museum, London. Located in a fashionable corner of London's South Kensington, the museum is a famous cathedral to nature, celebrating the extraordinary former diversity of life on Earth. Built between 1873 and

1880, the British architect Alfred Waterhouse (1830–1905) dreamt up the design and terracotta façade. Over 4 million visitors pass each year through the grand entrance set at the top of the broad sweeping stairway leading up to the main building. On reaching the top of steps, few visitors notice that Waterhouse decorated the Romanesque columns framing the doors with diamond-patterned stonework celebrating the fossil bark of *Lepidodendron* trees. These long-extinct giants of the Carboniferous coal-swamps photosynthesized with narrow strap-like microphyll leaves.

Palaeobotanists infer the existence of auxin transport, and even its direction of travel, from the graining patterns recorded in specimens of fossil wood from these ancient trees.[69] Although metabolized quickly, auxin leaves behind a lasting record of its action on the growth of trees. The direction of transport along the stem is diagnosed from the arrangement of the water-conducting cells (tracheids) that form the grain of the wood. When auxin is transported downwards from the apex of the tree with an undisrupted flow, the tracheids align end-to-end, creating the smooth grain of wood. When the flow of auxin is impeded by a disruption, such as a branch, it pools above the branch, causing tracheids to develop in swirls. Such distortions reflect changes in the direction of the flow of the auxin signal. Cut thin translucent sections of fossilized stems and roots of lycopsid trees from the Palaeozoic and you can observe, through a microscope, cellular disruptions in the dense matrix caused by tangled tracheids. These distortions are reminiscent of those caused by directional auxin transport in the wood of modern seed plants. Distorted vascular tissues in the trees that dominated the moist Carboniferous lowlands tell us that auxin regulated tree growth long before flowering plants appeared on the scene. But then the diamond patterning on the stone pillars of the museum entrance tells us this too. For auxin makes the diamonds, which are really scars left by falling leaves on the bark of the giant *Lepidodendron* trees.

Auxin is not the only plant hormone—plants require collections of them to manage different aspects of how they grow. Of all the challenges that living on land brings, surmounting the fundamental problem of water stress must have been key. This is where the hormone ethylene may have played a crucial role because it enables the 'submergence escape response' of plants. It works because ethylene diffuses more slowly in water than in air, leading to its entrapment and accumulation in submerged tissues. The accumulation of ethylene then triggers the escape response, which is essentially a growth spurt. The submergence of rice plants, for example, causes the ethylene escape response to kick in and trigger rapid stem growth to escape complete submergence. Not for nothing are two of the rice genes involved called *SNORKEL1* and *SNORKEL2*. The ethylene link to early land-plant survival is obvious. Having made the transition from an aquatic to a terrestrial habitat, land plants needed a mechanism for keeping their heads above water when the rains came. Proof that it appeared early on in the scheme of things came with experiments on bryophytes. If you subject mosses to submergence for a few days, genes in the ethylene-signalling pathway are activated, causing cells to elongate and shoots to extend, as mosses struggle through the surface meniscus to emerge above the water line.[70]

This escape pathway originates in charophyte algae, which possess collections of genes for the biosynthesis, transport, and perception of ethylene. Experimental work has now demonstrated that the cellular growth of the charophyte alga *Spirogyra* responds to ethylene and that a flowering plant (*Arabidopsis*) can use *Spirogyra*'s genes in its own ethylene-signalling pathway.[71] Ethylene, then, was likely a functional hormone in the common ancestor of freshwater algae and land plants that lived over 450 million years ago, and probably much longer ago than that. Interestingly, the ethylene receptor gene itself seems to be far more ancient, having been acquired by plants from cyanobacteria over a billion years ago. Chlorophytes had it and lost it. Charophytes had it and kept it.

Chlorophyte algae are evolutionary duds when it comes to giving rise to land plants, whereas charophyte algae, we are coming to appreciate, really are special. Why would the aquatic ancestors of land plants need an ethylene escape pathway? The answer to that question is unclear—unless we suppose charophytes were already living on land long before terrestrial plants evolved and got pushed back into freshwater by competition with the increasingly successful land plants to which they had given rise (the ignominy!).[72] If that is the case, the ethylene escape pathway could be regarded as a former trait, which they have not yet lost, from their days on land.

The fascinating story of ethylene shows how unravelling the history of a particular hormone shines new light on key events that laid the foundation of terrestrial plant life hundreds of millions of years ago. Our understanding of hormones in this context is undoubtedly limited, and largely confined to saying a few things about when particular genes originated and speculating about their function, but the powerful growth hormone gibberellin is an exception. Nick Harberd, also at Oxford as it happens, has dissected how gibberellin works, how the genetic machinery arose, and how it benefited early land plants.[73] Gibberellin is named after the fungal pathogen *Gibberella fujikuroi* (now called *Fusarium moniliforme*), the effects of which were first recognized in Japan in 1935. Rice seedlings infected with the fungus develop a disease known as 'foolish seedling' syndrome,[74] caused by the fungus secreting gibberellin. The symptoms include the rapid and excessive elongation of rice seedlings with slender leaves, and drastic reductions in rice yields.[75] Why the fungus bothers to inflict these traumas on the rice plant by secreting gibberellin only became clear when it was demonstrated that the gibberellin disables the plant's immune system, allowing greater infection by the fungus and boosting its virulence as a result.[76]

Harberd elucidated the mechanism by which gibberellin stimulates the growth of flowering plants and it is worth unpacking some of the detail to see how these DNA-based instructions had far-reaching consequences for the evolution of terrestrial plant life.[77] His team discovered that gibberellin allows growth by triggering the destruction of certain protein molecules, transcription factors called DELLAs. DELLAs are restraining orders for plants; anti-social behaviour orders for shoots, preventing riotous growth.[78] Gibberellin binds to a receptor molecule to form a chemical complex inside the cell's nucleus. The details are complicated, but in short this stimulates the tagging of DELLAs with a small but crucially important protein ('ubiquitin'). These reusable protein tags are the kiss of death for DELLAs, labelling them for destruction by the waste disposal units of cells, proteasomes.

The biochemical tagging of proteins through the attachment of special molecules such as ubiquitin is actually a general mechanism by which cells destroy unwanted proteins. Cells of all living organisms use this method of protein destruction, which goes by the unwieldy name of '*ubiquitin-mediated protein degradation*', and it won the three scientists who made the ground-breaking discovery (Aaron Ciechanover, Avram Hershko, and Irwin Rose) the 2004 Nobel Prize for Chemistry.

In short, gibberellin causes DELLAs to disappear, and this removes a molecular restraint on plant development and allows the growth-promoting genes to get on with the business of actively growing the shoot. The molecular machinery involved in plants is very ancient, with the full set of genetic instructions for regulating the architectural exuberance of shoots assembled in distinct stages over the past 400 million years.[79] By the time seed plants had evolved, the complete genetic toolkit had appeared, and, because environmental cues modulate DELLA activity, plants now had a mechanism for coping with changing environmental conditions. When it is too hot or too cold, or the soil dries out during a drought, DELLA activity changes, telling the plant when it is time to slow down. When conditions improve, DELLA activity changes again and the plant 'knows' to resume growth.[80] It provides a means of guiding plants successfully through the delicate game of life on land by ensuring their survival during times of environmental crisis. In Darwinian terms, this speaks directly to the 'survival of fittest'.

If these intricate molecular details of the goings on inside plant cells seem rather esoteric and far removed from reality, nothing could be farther from the truth. They have real world importance in explaining the Green Revolution of the 1960s, led by Norman Borlaug (1914–2009), the 'Father of the Green Revolution', who received the Nobel Peace Prize in 1970 in recognition of his contributions to world peace through increasing food supply. Selective breeding of crop plants for dwarfing traits brought about the spectacular increases in wheat and rice yields that fuelled the 'Green Revolution'. Before the Green Revolution, older varieties of wheat and rice plants had tall, weak stems unable to support the heavy package of grains loaded at the top of slim stalks, causing the plants to fall over in the wind. Lodging, as the process is known, reduces yields because the grain in the head of the stems is ruined on the flattened plants. Compact dwarf varieties of wheat and rice overcame the problem, with stout stalks easily supporting the grain-packed heads and therefore being far more resistant to lodging by wind. After farmers adopted these 'dwarf' varieties, wheat yields in Europe, North America, and Asia soared. This confounded the dire predictions of the Stanford ecologist Paul R. Ehrlich, who forecast in his bestseller *The Population Bomb* (1968) that 'The battle to feed all of humanity is over…In the 1970s and 1980s hundreds of millions of people will starve to death in spite of any crash programs embarked upon now'.

The Green Revolution was ushered into the molecular era with the identification of the genes responsible for the new dwarf varieties of wheat and rice. As you might have guessed, alterations to the DELLA genetic toolkit were at the heart of it.[81] Wheat varieties producing stout ramrod-straight stalks had mutated DELLA genes, which prevented gibberellin from relieving their growth-repressing effects, causing the crops to become stunted. Rice varieties meanwhile had an impaired ability to synthesize gibberellin itself, meaning DELLA-opposed growth predominated, resulting in shorter plants. So DELLAs, along with irrigation, chemical fertilizers, and agro-chemicals, help feed the world today and have been underwriting the solar-powered economy of nature since before the dawn of terrestrial plant life.

Deep insights into the origin and diversification of the land plants, woven into the DNA of our modern terrestrial floras, are starting to tell us about wonderful long-hidden molecular events that facilitated the colonization of the land. We are edging towards a deeper understanding of the continued metamorphosis of plant life as it evolved from simple freshwater algae to complex trees. The Devonian 'big bang', fuelled by the sorts of genetic toolkits regulating development programmes we have been considering, was really a Palaeozoic revolution, with plant life continually reinventing itself. Variety in plant form across the Devonian Period was generated through the assembly and rewiring of genetic toolkits from pre-existing genetic components. Evolution operated by working with what is to hand to generate novelty. Similar modes of evolution enabled the evolutionary metamorphoses of animals.[82] Expansion and rewriting of genetic toolkits is explaining how insects evolved wings, how beetles grew horns to fight over females, and how moths and butterflies decorated their wings with brightly coloured eye-spots to ward off predators.[83]

The famous Oxford mathematician G.H. Hardy, who held the Savilian Professorship of Geometry there between 1919 and 1931, gave biologists the Hardy–Weinberg principle describing gene flows through populations of plants and animals. He remarked in his classic essay *A Mathematician's Apology* (1940) that mathematicians probably have the best chance of achieving immortality, because of the fundamental nature of the mathematical proofs they produce. Hardy thought this 'very comforting for dons, and especially for professors of mathematics'. He was writing at a time before we understood the secret of plants' success in colonizing Devonian landscapes. It is the same as their ability to defy ageing and is found by looking at plants in botanic gardens. Their modular construction is controlled by ancient networks of genes that permit organs like leaves and roots to be repeatedly renewed as they show signs of physical deterioration with age and attack from pathogens.[84] It also helps them avoid the gradual accumulation of harmful mutations, as scientists showed when they sequenced the genome of 'Napoleon's oak' (*Quercus robur*) on the campus of the University of Lausanne, Switzerland.[85] The tree is so-called because it was planted as a young individual (22 years old) in 1800 to commemorate the visit of the general and his troops on their way to conquer Italy. The mutation rate in this 234-year-old tree was 10 times lower than in the short-lived weedy annual *Arabidopsis*. How can this be for such an ancient tree? The answer is, of course, that the clusters of stem cells on the growing tips on the branches develop their own stems, leaves, and shoots over and over again. Unlike humans, trees slow down ageing through renewal, or, as Philip Larkin puts it in 'The Trees', 'Their yearly trick of looking new'.[86] The modular growth habit of trees means that harmful mutations may cause leaves or branches to die off but will not kill off the whole tree.

The concept of plant growth based on repeated modules of shoots and roots perplexed Charles Darwin. He wrote in his *Origin of Species* (1859) that:

> All organic beings have been formed on two great laws – Unity of Type and the Conditions of Existence. By unity of type is meant that fundamental agreement in structure, which we see in organic beings of the same class, and which is quite independent of their habits of life. The conditions of existence are fully embraced by the principle of natural selection [which] acts by either now adapting the varying parts of each being to its organic conditions of life; or by having adapted them in long-past periods of time.

What Darwin really meant by his phrase 'Unity of Type' was that the body plan of all plants is the same, conforming to a modular construction consisting of the three vegetative organs we have been considering throughout this chapter: roots, shoots, and leaves. This common body plan is repeated throughout the plant kingdom, but with infinite variation in the details. What puzzled Darwin about this view of the world was the thorny issue of the mechanism that simultaneously allowed plants to retain a common body plan, and yet gave rise to the enormous complexity of plant life we see today.[87] The question has intrigued scientists for more than 150 years. We are now starting to find some answers: flexible green genetic machinery generates diversity in the modules—leaves, flowers, roots, and so on—while allowing plants to stick with the same successful basic body plan.

Beyond Hardy's window, in the College grounds, 'the trees are coming into leaf' with the arrival of the warm spring sunshine, heralded by the voices of Magdalen College choir. And, it is Larkin again in 'The Trees' who puts it so well, 'Yet still the unresting castles thresh/In full grown thickness every May': each leaf unfurls studded with thousands of tiny mouths; each mouth is a sophisticated microscopic gas valve for swapping gases with the atmosphere. The gas valves of modern trees are remarkably similar to those that first adorned the young naked shoots of early vascular land plants over 410 million years ago. These structures have not yet got much of a mention, but proved to be one the most important innovations in the kingdom of plants. They ushered in a new mode of plant life that fundamentally changed the cycling of elements through the biosphere and radically transformed the planet. In Chapter Five, we discover how.

# 5 GAS VALVES

'Plants are like "beautiful strangers"—they are so different from us. They don't have neurons or brains, yet they cleverly sense the surrounding environment, defend themselves and prosper.'

Keiko Torii, 2013, *Current Biology*, 23, R943–4

Walking through the groves of giant redwoods (*Sequoiadendron giganteum*) in the Yosemite National Park gives you the ground-up perspective of these magnificent towering trees. Giant redwoods are the largest trees on Earth, with girths wide enough to drive a Cadillac through, should you choose to, but they start life as tiny seeds measuring about a millimetre long. Gaze upwards from the darkness of the forest floor and the enormity of the trees as they point towards the sky, saluting the Sun, is immediately obvious. Beneath these leafy cathedrals is a tranquil sanctuary from the modern world created by the ancientness of trees over millennia. Viewed from the air, a quite different perspective becomes apparent. Scattered across the ridges and valleys of the western slopes of the Sierra Nevada Mountains, California, natural groves of these huge trees are dotted over thousands of hectares. Well known to Native American tribes living in the region, long before their discovery by European settlers early in the nineteenth century, these massive photosynthetic organisms remind us of the extent to which plants have truly conquered the business of living in thin air.

Understandably taken with these majestic trees, the producers of the BBC Television series *How to Grow a Planet* opened each episode with a filming sequence showing dramatic sweeping panoramic views of the Sierra Nevada forests. The series provided a companion to its sister series, *Botany: A Blooming History*, presented by Timothy Walker, former Director of the Botanic Garden at the University of Oxford. In mischievously highlighting these commendable contributions by the BBC to the fight against plant blindness, I should disclose that the former series was based on my earlier 2007 book, *The Emerald Planet*. Both the book and TV series hoped to offer readers and viewers an extraordinary new perspective on the role of plant life in shaping Earth's history. History will judge whether this was the case—the point is that each episode showing Sierra Nevada's rocky landscape and its *Sequoiadendron* trees opened with the camera tightly focussed on the ebullient Scottish presenter Iain Stewart. Only when it zooms out do we discover Stewart roped to the top of a giant redwood tree, where he proclaims the power of the plant kingdom in transforming the Earth over millions of years.

Yet an overlooked secret of the success of redwood trees only becomes evident by scrutinizing the fine details. Zoom in close, past Iain Stewart, past the branches, past the feathery leaflets, and focus down onto the cellular details of a single needle. There, on the surface of each needle, are special elongated cells, lined up like soldiers flanking the central midrib from tip to base. These microscopic soldiers are paying close attention to the atmospheric conditions and receiving commands, via hormones, from roots sensing the moisture of the soil, not to mention the electrical currents signalling changes in the water status of the needles themselves. Each soldier comprises a pair of specialized guard cells that flank an adjustable stomatal aperture or pore (from the Greek *stoma* for mouth). As weather conditions change, or the soil dries out, the tiny pores respond by opening and closing, tightly regulating the molecular exchange of gases between needle and the surrounding atmosphere to limit excessive water loss from the battalion of cells that make up the rest of the leaflet.

The desiccation of delicate tissues is an ever-present danger for plants because in the essential business of absorbing carbon dioxide from the atmosphere for photosynthesis, leaves must open their stomatal pores. As they do so, water immediately escapes, evaporating from the wet surfaces of cells inside the leaf and streaming out into the drier surrounding atmosphere. In fact, carbon dioxide molecules have to enter the leaf against this outward flux of water molecules—the transpiration stream. Mainly as a consequence of opening pores and exposing the wet cell surfaces of the leaf interior to a drier atmosphere, it costs plants about 1 kilogram of water to synthesize a few grams of plant tissue from carbon dioxide. The actual amount depends on weather conditions, especially atmospheric dryness, and the photosynthetic mode of the plants.

For trees, and indeed virtually all terrestrial plant life, water is the price to be paid for the freedom of living in thin air. Yet the dilemma facing land plants is that water is also a precious commodity, essential both for running their metabolism and for keeping them upright by pressurizing cells to give tissues rigidity and drive cell expansion for growth. The pressures involved are enormous. Encased by rigid cell walls, plant cells withstand pressures three times those found inside a bottle of champagne. Not surprisingly, blind, ruthless evolutionary trial and error has engineered plants with exquisite capabilities for managing both sides of their watery economics—its supply and demand—as they refuel with carbon dioxide. Roots supply water to the tree; water-conducting tissues in the trunk and branches deliver it to leaves; and stomata trade its evaporative escape for carbon dioxide in the glare of the Sun.

Consider the harsh realities of the situation facing early plant life when it began adapting to a terrestrial existence half a billion years ago. Survival depended on evolving a strategy balancing the imperative to grow against the risk of death by desiccation. Secreting a waxy impermeable outer coating, a cuticle, surrounding the delicate shoots thrust upwards into the air provided part of the solution. Indeed, a lipid-rich cuticle may have been inherited from their freshwater algal ancestors, who needed it to avoid drying out when falling water levels left them stranded on muddy banks.[1] But the impervious cuticle itself created a problem: it impeded the ability of simple early land plants to absorb carbon dioxide. Specialized miniature gas valves with adjustable apertures—stomata—provided the solution to the permeability problem created by the cuticle. These remarkable cellular innovations were actually sophisticated microscopic pores that proved pivotal to plant life as it expanded across the landscape to transform the planet's continents.

The oldest fossil stomata unearthed to date are those from 418-million-year-old Silurian rocks in Wales.[2] They belong to the iconic fossilized land plant *Cooksonia*, the earliest known simple vascular land plant, which grew in clusters of slender leafless stems the width of a needle, if that, and stood only a centimetre or so tall (**Figure 12**). View *Cooksonia* fossils under a scanning electron microscope and the wonderful detail of ancient stomata is visible, created by a pair of specialized kidney-shaped guard cells that closely resemble those on the leaves of modern plants. Indeed, for all early vascular land plants, the presence of stomata is the rule not the exception. Microscopic examination of their tiny fossilized remains, dating to the Silurian from localities worldwide, confirms that stomata are dotted across the surfaces of their stems and reproductive structures (sporangia) held on the tips of slender stems.[3] Written in stone, the anatomy of these fossils is testament to the central role stomata have played in regulating the fundamental plant processes of photosynthesis and transpiration for hundreds of millions of years.

By performing this important function, stomata, coupled with other innovations we have already considered, including rhizoids to absorb nutrients and water

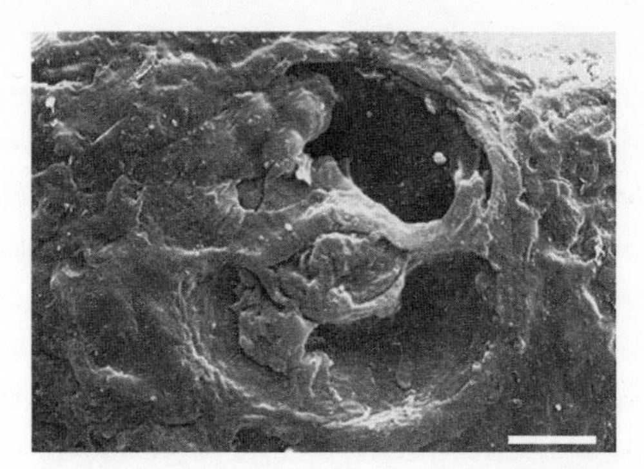

**Figure 12** A scanning electron micrograph image of an early Devonianian vascular plant's (*Cooksonia*) 410-million-year-old stoma composed of two kidney-shaped guard cells, much like those possessed by all living land plants (scale bar = 10 μm).

from the soils, and vascular tissues for transporting water through stems, proved vital in allowing Earth's early land floras to breathe deeply and grow taller. The tallest trees on Earth, the Californian coastal redwoods (*Sequoia sempervirens*)—close cousins of the giant redwoods—illustrate how this may have happened long ago.[4] Shackled to the flimsiest branches of redwoods in the Humboldt Redwoods State Park, north of San Francisco, and braving the high winds, intrepid plant physiologists have revealed the secret role of stomata in permitting these trees to attain great heights of over 100 m (**Plate 4**).

In these tall trees, water escaping through the stomata at the treetops generates an enormous transpiration stream; the canopy of each tree loses hundreds of kilograms of water every day.[5] This loss of water, the 'cost' of photosynthesizing on land, solves its own problem of supply. It helps 'pull' water into the rooting systems from the soil, through the hollow water-conducting cells of the trunks, to the very tops of the trees, to overcome the opposing forces of gravity and friction. The water-conducting cells themselves are long, hollow tubes stacked vertically through the stem, often with thickened cell walls for reinforcement. The transpirational 'pull' of water, as the process is known, tends to be greatest near the top of the tree, where water is evaporating through stomata.[6] In coastal redwoods, water moves upwards at a stately pace, taking several weeks to travel from the base of the trunk to the top of the tall trees, where it pressurizes cells and drives the growth of leaves by cell expansion. We can think of the continuous column of water from roots to the treetop as being under tension between these opposing forces of gravity and transpiration, and the column holds together because of the adhesion of water to the walls of water-conducting cells and the cohesive bonds attracting water molecules to each other.[7] This is the so-called cohesion-tension theory and since its introduction in the late nineteenth century, it has become widely accepted as the mechanism explaining the rise of water in plants.

The idea that 'strings' of water are lifted up through the plant by its evaporation from leaf surfaces is extraordinary, and it originated over a century ago in 1895. When first proposed by Henry H. Dixon and his colleague J. Joly at Trinity College, Dublin, Francis Darwin (1848–1925), Charles Darwin's third son, who had a great passion for stomata, was incredulous. Darwin commented, 'To believe that columns of water should hang in the tracheals [water-conducting cells] like solid bodies, and should, like them, transmit downwards the pull exerted on them at their upper ends by the transpiring leaves, is to some of us equivalent to believing in ropes of sand'.[8] Sir Isaac Newton (1643–1727), on the other hand, would presumably have been delighted with the idea. Newton penned his perceptive ideas about how light knocks away fluid particles escaping from the shoot in an obscure undergraduate notebook sometime between 1661 and 1665.[9] He wrote that 'by this meanes [*sic*] juices continually arise up from the roots of trees upwards'. So the father of gravity himself thought of an explanation for how plants oppose gravity that has more than passing resemblance to our modern understanding. And he did so two centuries before botanists suggested the widely accepted cohesion-tension theory explaining the ability of plants, from grasses to the Earth's tallest trees, to transport water from roots to leaves.[10] As Emily Conover of the international journal *Science* snappily remarked in commenting on the matter, 'Although he didn't quite get the details right, one thing's for sure—Newton was no sap'.[11]

The tussle between Newton's gravity and transpiration means that trees face a problem as they grow taller. With increasing height, the effectiveness of transpiration in overcoming the forces of gravity and friction diminishes, and this reduces the flow of water into the cells of leaves, making it difficult for them to expand, and slowing their growth. Those cells at the top of trees are very small and produce thick, dense leaves that are less effective at photosynthesis, limiting the maximum height of the trees.[12] The coastal redwood trees in the Humboldt Redwoods State Park stand some 113 m tall and are over 2000 years old, and still have some way to go before topping out at around 120–130 m. Reaching those heights could take another 30 or 40 years. The paradox of how these trees, losing hundreds of kilograms of water every day, persist in the Californian climate with a long dry season is resolved by the realization that during droughts they switch to the substantial reservoir of water stored in the outer few centimetres of their trunks. The tree canopies also intercept and capture fog drifting in off the Pacific Ocean. As it condenses on the foliage, water droplets collect and drip onto the soil to provide another source of water that helps tide them over until the rains arrive.[13]

We can now begin to see how the collective activities of tiny gas valves play a crucial role in allowing *Sequoia* trees to grow tall by generating a transpiration stream pulling water into roots from the soil and through their trunks to the treetops. And think about how the colonnades of redwoods sprouting plumes of green foliage rising tens of metres into the air, peppered with stomata, achieve this feat. Plants cleverly harness the power of the sunlight to drive evaporation and generate transpiration, without expending extra energy. It's a similar story for all plants that make up the great diversity of Earth's land floras. The packaging and longevity of the plants involved differs, but the fundamental physics of the processes remains the same.

Now consider the alternative situation—that facing the earliest plants making the first tentative move to land without stomata (the proto-bryophytes, Chapter Two). Living on thin soils with a limited ability to soak up and retain rainwater, perhaps similar to those found in the Mojave Desert today, and attached by simple rhizoids, photosynthesis in these pioneering plants could only get going after it rained. It could only continue provided the surface of simple soils on rock surfaces remained sufficiently moist to permit capillary forces to draw up water. This wicking process requires a continuous film of liquid water connecting one particle of soil to the next in an unbroken chain to sustain the flow, and is driven by evaporation on the surface by sunlight. The problem is that as water evaporates from the soil surface, air enters the larger pores between particles lower down, gradually breaking the continuity of the water film. Once the connection is broken, it creates a roadblock, preventing the shallow rhizoids from accessing liquid water deeper in the soil profile.[14] Under such challenging conditions, early land plants lacking stomata would have been unable to regulate water loss and unable to access soil water. These plants would have had no choice but to remain small, enter dormancy, and await the next rains, like many bryophytes do today, to avoid desiccation.

The appearance of plants with stomata, on the other hand, marked the beginning of a radical new mode of plant life.[15] Thanks to this cellular innovation, plants could sensitively control evaporative water loss from their shoots and pull water into their roots from soils. This meant they could access water deeper in the soil profile, and they gained the ability to survive on soils that dried out between rainfall events. It sounds like such a simple thing, but it proved revolutionary, because plant life had established a continuous watery connection between the chloroplasts doing photosynthesis and roots in the soil. Plants could now stay hydrated between rainfall events by regulating water loss and avoid going into dormancy. Success, of course, hinged on the clever decision-making of stomata that help keep the watery linkages intact while negotiating the optimal exchange of water for the carbon dioxide needed to fuel photosynthesis.[16]

As land floras composed of stomata-bearing plants spread across the continents, the burgeoning terrestrial biosphere began exerting a growing influence on how water cycles around the Earth. For plants are really upside-down waterfalls, showering the atmosphere with water vapour that eventually returns to Earth as rain. At the global scale, the stomatal transpiration of modern forests, grasslands, and crops releases an estimated 32 billion tonnes of water vapour into the atmosphere each year, double the water vapour content of the atmosphere from other sources. This is enough to play a significant role in the global water cycle[17] and we should probably not be surprised to discover that roughly 40% of the rainfall on land originates from the transpiring canopies of modern forests.[18] In a world without plants enriching the atmosphere with water vapour, the climate would be drier, and the landscape more barren. But as plant life cloaked the land surface in green photosynthetic leaves perforated with pores, it gradually began re-tuning the hydrological cycle, capturing an ever-increasing amount of precipitation from the soil and recycling it back into the atmosphere as transpiration. Simulations with global climate models illustrate how different things might be without plants.[19] In one simulation, a fully vegetated Earth ('green world') actively transpires water vapour, and in the other simulation transpiring vegetation is removed ('desert world'). Compare the two simulations and an idealized picture of the role plants play in shaping global climate emerges. In 'desert world', rainfall over continental interiors dries up completely and summer temperatures are nearly 20°C warmer than in 'green world'. In other regions, land areas proved less sensitive to the absence of vegetation, especially around the margins of continents and in the tropics, where abundant precipitation fell even without plants. Could these areas offer clues to where plants began to make landfall?

Whatever the answer to that question, we can speculate that by adding water vapour to the overlying atmosphere, diversifying floras slowly created conditions more hospitable to facilitating their continued colonization of the land.

## GAS VALVES RELEASE THE PRESSURE ON DEVONIAN FISHES

Plant life equipped with stomata and its revolutionary watery linkages between roots and shoots soon spread into a new world of light and air, with large trees and forests taking hold on the continents. The expanding biomass of these terrestrial floras enhanced the burial of organic material on land through the Silurian and Devonian, and thereby gave Earth a huge new source of oxygen. Oxygen added to the atmosphere by photosynthesis is normally used up by microbial decomposition, balancing the budget. But if carbon burial occurs before decomposition, the oxygen instead feeds into the oceans and atmosphere. Through these processes, the evolutionary assembly of stomata-bearing terrestrial floras breathed oxygen into the Devonian world.

Chemical analyses of sediment cores drilled from the sea floor reveal the telltale rise in seawater oxygen levels through the Devonian.[20] At the same time, the Devonian oceans witnessed a dramatic increase in the maximum size of large predatory fish. Before the Devonian oxygenation event, fossil fish of Ordovician and Silurian times reached lengths of a few tens of centimetres. Afterwards, fish greater than a metre long appear in the fossil record. Are these changes in the fossil record of fish and ancient atmospheric oxygen levels merely coincidence, or could they be linked?

The short intense bursts of speed modern fish need to capture prey entail a high demand for oxygen to fuel their muscles. The same principle also holds for large Devonian predatory fish, with the implication that the evolutionary arrival of these finned swimmers with energetic lifestyles in the oceans was only possible after oxygenation of the biosphere.[21] It could also be that this development reflected an expanded food web supported by nutrient-rich runoff flushed into the oceans from plant-covered continents. Still, it seems the rise of stomata-bearing plants played a crucial role in driving a critical transition in Earth history—the development of a fully oxygenated Devonian Earth system that ushered in the Devonian 'age of the fishes'.

Because as a green, chlorophyllous Eden took hold, plants inadvertently set up a series of self-reinforcing (positive) feedback processes that generated a climate increasingly agreeable for the further spread of plant life. More plant life on the continents meant the capture of more precipitation by roots and its return to the atmosphere to seed rainfall (**Figure 13**). By adding water vapour to the atmosphere, plants not only cooled the land surface (as sweating cools your skin) but could also have increased the cloudiness of the sky, casting additional cooling shade. Together, these effects generated convective activity in the atmosphere, pulling moist air in from the ocean onto the land, further enhancing precipitation.[22]

Regional-scale meteorological effects like this partly explain why precipitation doubled over land in the 'green world' versus 'desert world' model simulations mentioned earlier. They also explain how the collective actions of stomata strengthened over time to air-condition the planet, keeping it a few degrees cooler, and the atmosphere moister, than it would otherwise be without plants.

Of course, there is no grand plan, no teleology about this. The collective actions of stomata on climate over time are simply a by-product of the crucial decisions they make on a minute-by-minute basis about when it is safe to open and close, by how much, and for how long. To do this, they operate as miniature sensors organized into fantastically complex arrays that integrate multiple cues from the environment, and chemical and electrical signals communicated from cells in leaves and roots. Scientists bewitched by investigating details of stomatal function agree that these marvellous cellular structures seem to operate continually in just such a way as to maximize the amount of carbon dioxide captured by the leaf while simultaneously minimizing water escaping as transpiration. Yet they remain baffled as to how stomata achieve this amazing feat.

The late Fred Sack (1947–2015), a giant in the world of stomatal biology, once likened it to Dr Pangloss's thoughts in Voltaire's 1759 French satire *Candide*. Candide, remember, is the illegitimate nephew of a German baron who grows up in the baron's castle under the tutelage of the scholar Pangloss. Sack remarked that 'life on Earth depends in no small part on stomata' and noted that stomata offered what Pangloss called 'the best of all possible worlds' for making the planet green and lush.[23] Joe Berry, a professor at the Carnegie Institution, Stanford, and another scientist who has also spent his career thinking about stomata, puts it another way,

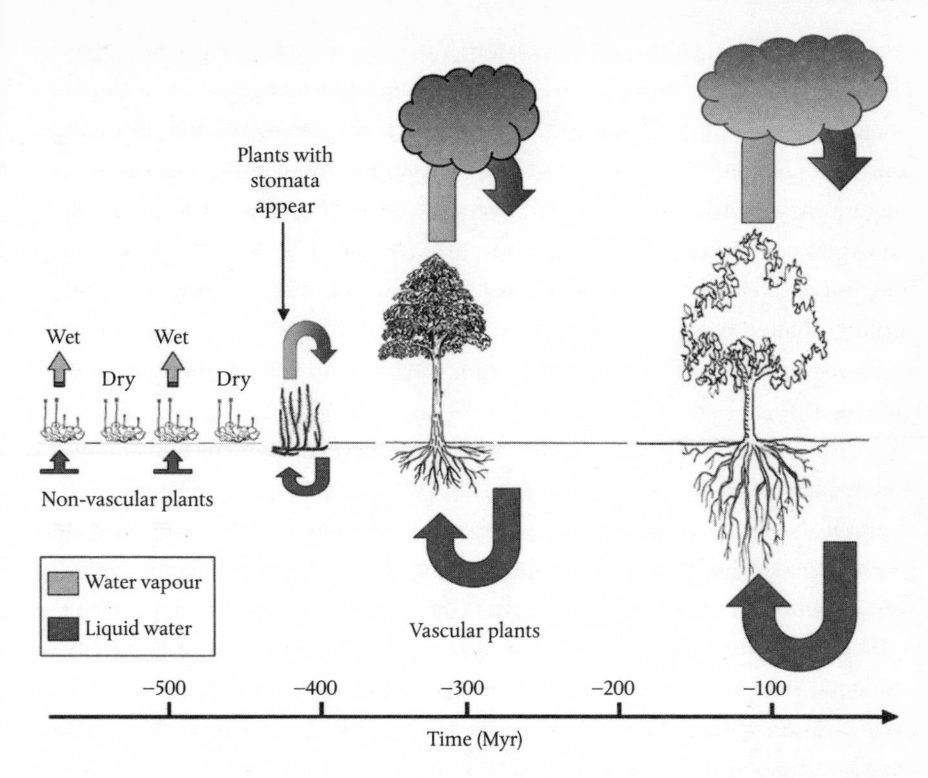


**Figure 13** The evolutionary progression from small plants lacking stomata to large stomata-bearing trees accelerated the global water cycle, recycling soil water by adding it as transpiration to the atmosphere to later fall as rain.

commenting, 'You might think of them as analogous to traders on Wall Street making bets on the cost of water (the currency) for carbon dioxide uptake (a commodity)'.[24] This useful way of framing the concept extends readily to the global scale. Just as the activities of Wall Street traders influence the world's economies, so the collective decisions of stomata influence global water and carbon cycling, and climate.

Today, we understand the basic mechanism of how pressure-operated guard cells work. In broad terms, molecular signals adjust the pore size by causing them to swell or shrink, although much of the fine detail remains a mystery. Molecular pumps embedded into the cell membranes drive pressure changes in the kidney-shaped guard cells (**Figure 14**). When environmental conditions are right, usually in the morning and when water is plentiful, stomata open by expending energy to drive molecular pumps that expel positively charged hydrogen ions out of the cells and create a minute electrical charge across the cell membrane. This charge opens the electronically controlled molecular gates of membrane-spanning channels to allow potassium ions to flood into the cell. In effect, the inside of the guard cells becomes saltier and this causes an influx of water by osmosis, pressurizing and inflating the cell, much like pumping up a bicycle tyre. The inner surfaces of guard cells are thickened, and bend apart in semi-circular fashion as they inflate, creating a pore. This arrangement provides an inherent fail-safe mechanism protecting the leaf against dehydration during severe drought; as the soil dries, wilting due to decreased water pressure automatically closes stomata and conserves the plant's remaining water reserves.

As we can start to appreciate, from evolutionary, biophysical, and physiological perspectives—take your pick—stomatal biology seems to have a unique complexity about it, with far-reaching consequences for all life on Earth, not to mention the global cycling of elements and energy around the planet. A key question follows: how did stomata originate? The fossil record of plant life suggests they turn up ready-made, apparently without earlier prototypes. Is it possible plants could

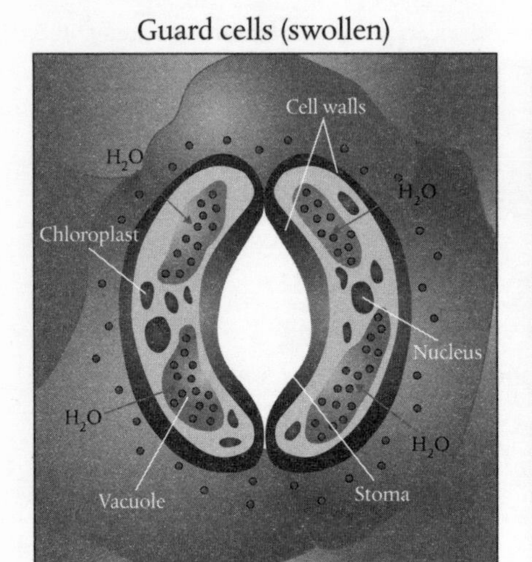


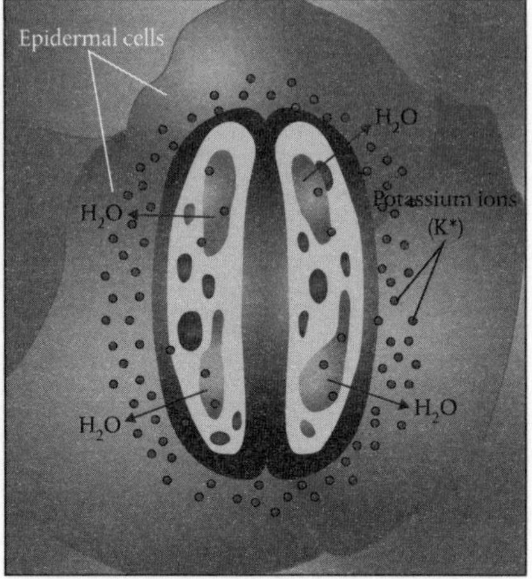


**Figure 14** Opening and closing of stomata.

## BEWITCHING STOMATA

The sophistication of these microscopic structures often casts a bewitching spell on the scientists who study them. Nearly two centuries ago, Erasmus Darwin (1731–1802), Charles Darwin's grandfather, suffered an obsession with stomata brought on by suffocating the leaves of Portuguese laurel and balsams by pouring oil over them; leaves treated in this way died within a day or two. The same thing happened when he subjected insects to similar treatments. Given insects breathe through tiny pores (spiracles) punctuating their thoracic segments, Darwin drew the conclusion that there was a 'similitude between the lungs of animals and the leaves of vegetables', both being used for breathing, without him actually observing stomata. The great botanist Sir Joseph Banks (1743–1820) had seen them, though, on the leaves of corn and wheat, noting stems were 'furnished with rows of pores... shut in dry and open in wet weather'. Banks was concerned with the cause of mildew on crops and supposed that infection took place through the stomata, which is generally correct.

Long before the days of Darwin and Banks these little structures had caught the attention of the great Italian polymath Marcello Malpighi (1628–1694).[25] Malpighi seems to have been the first person to observe stomata through a microscope, and later published drawings of them, but Darwin and Banks were probably unaware of them. I should also mention that the English anatomist Nehemiah Grew (1641–1712)[26] ran him close with what he described as 'breathing holes' on the surfaces of leaves in 1682. Malpighi's investigations were not confined to the anatomy of plants and several microscopic anatomical structures of the animal world are named after him. There are the Malpighi skin layer and two different Malpighian corpuscles in the kidneys and the spleen, as well as the Malpighian tubules in the excretory system of insects. It is probably only an accident of history that stomata are not actually called Malpighian mouths, or some such thing.

have evolved these sophisticated structures in such a fashion to learn how to breathe? What sort of genetic toolkits did early land plants like *Cooksonia* and its cousins deploy to build them? As with questions on the origins of leaves and roots, we have to delve into the world of evo-devo to see further than the fossil evidence.

Mosses are an important lineage for these sorts of investigations because they are among the oldest living group of land plants with stomata (Chapter Two). They diverged from a common ancestral lineage over 400 million years before flowering plants appeared on the scene. The logic is that if mosses and flowering plants share similar genetic mechanisms for building stomata, then we can infer that they already existed in the last common ancestor of these plants, and this takes us a significant step closer to *Cooksonia*'s stomatal genes.

To develop the evolutionary story, we have to start with what is known about the business of building stomata in flowering plants. And, in particular, with the breakthrough discoveries made by Dominique Bergmann[27] and Keiko Torii[28] concerning three 'master regulator' genes that sit at the heart of the genetic machinery for making stomata. Both are world-leading molecular biologists: Bergmann is a distinguished professor at Stanford University, where she is a colleague of Joe Berry's and where a preoccupation with stomata is considered quite normal; and Torii is a distinguished professor at the University of Washington, Seattle. For reasons that will become clear shortly, the three major genes they discovered that guide stomatal development are called *SPEECHLESS (SPCH)*, *MUTE*, and *FAMA*, with Bergmann's lab discovering *FAMA*[29] and Bergmann and Torii's labs then separately and yet simultaneously identifying the other two.[30]

Together, these genes shape the molecular pathways regulating the processes of cell division and cell fate by which plants produce guard cells. The cellular process of making stomata begins with the simple act of cell division that produces two daughter cells of different size. This ability to undergo what is essentially an asymmetric cell division should not to be underestimated. It sounds like a trivial thing, but it is actually the key to making all sorts of specialized structures in plants and animals because it generates two cells with potentially different fates, leading to the formation of distinct cell types and new organs.[31] In stomatal development, the smaller cell gains special, distinctive, stem-cell-like properties and undergoes several further divisions. Next, the cell at the centre of the rosette slowly matures, taking on an oval shape, and transitions into a cell that will eventually undergo a symmetric division to form a pair of specialized guard cells.

The role of the newfound genes in regulating these processes of cellular differentiation is well established, and it is worth unpacking a little of the detail to appreciate how it sets the stage for discoveries with the moss genetic toolkit (**Figure 15**). *SPCH* controls the first step in making stomata. In the young leaf, cells that express *SPCH* look the same as those that do not, but they are doing something unique. They are preparing to undergo the asymmetric cell divisions that will initiate stomatal development. In *Arabidopsis* plants lacking *SPCH*, leaves are unable to develop stomata, stunting growth and ultimately causing them to die. *SPCH* does not just act once, but is kept switched on in the smaller daughters of the asymmetric divisions, and this can amplify the number of cells in the epidermis. These cells are on their way to becoming stomata, but they have not yet committed themselves to that outcome. Commitment is the role of the next gene, *MUTE*. A clue to its role came when geneticists created *Arabidopsis* lines lacking *MUTE* and found that the leaf surfaces were covered in rosettes of cells (from asymmetric cell divisions), but no stomata had been made. *Arabidopsis* plants with *MUTE* switched on all the time produces the reverse effect—leaves covered with stomata. *FAMA* regulates the final stages of stomatal development. Named after the many 'false mouths' of the Roman goddess of rumour (Fama), this gene is specifically expressed in immature guard cells. *Arabidopsis* plants lacking *FAMA* develop leaves with stacks of cells aligned horizontally on the surface that resemble segmented caterpillars. These arise from repeated divisions of

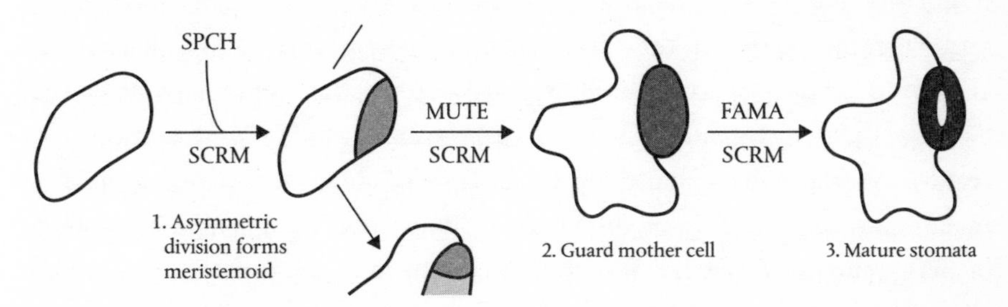


**Figure 15** Three master genes (*SPCH/MUTE/FAMA*) control cell division and cell fate to make stomata on the leaves of flowering plants through interaction with *SCRM*.

proto-guard cells as they attempt but fail to complete the process of stomatal development. Without *FAMA*, leaves are unable to close out the final step in building stomata. We can see now that each of the three genes, activated in discrete places, and at particular times, performs a specialized function to ensure the normal process of stomatal development. Other genes are involved in generating the diversity of shapes and sizes we see in leaves across the plant kingdom, but this core set of three master switches is all that need concern us here.

Our next task is to see how these findings fit with recently discovered genetic mechanisms guiding the formation of stomata around the base of the little spherical capsules (sporophytes) of the moss *Physcomitrella*. In fact, this moss has two stomatal genes closely related to the *SPCH/MUTE/FAMA* genes of *Arabidopsis*,[32] called *SMF1* and *SMF2*.[33] Until recently, we had little idea what roles either gene played in moss stomatal development, if any, but then, over a period of about a decade, an international collaborative team of researchers, including researchers from Sheffield, Freiburg, Stanford, and Leeds, began to unpick the mystery.

The long journey to discovery began when we took the obvious step of cancelling millions of years of evolution and manipulating the genetic make-up of moss by deleting either *SMF1* or *SMF2* to investigate what happened. To our delight, the resulting lines of moss missing a single gene (*SMF1*) developed normal-looking sporophytes in every respect except that they lacked the characteristic ring of stomata around the base of each capsule (**Plate 5**). It was a 'eureka' moment for Caspar Chater, the dedicated researcher who had started working on the project as part of his PhD a few years earlier. His time ran out before he obtained definitive results, but he later returned to doggedly pursue it as a postdoctoral researcher with us in Sheffield. On first discovering the 'missing stomata', Chater took a moment to collect himself before anxiously examining other moss sporophytes under the microscope. In the event, he need not have worried. His observations confirmed a beautiful anatomic anomaly: the sporophytes of mosses lacking the *SMF1* gene lacked stomata. In contrast, mosses lacking *SMF2* developed with stomata as normal. We can conclude, then, that *SMF1* but not *SMF2* is essential for making stomata in moss, and this fits neatly with earlier complementary findings by Bergmann's group. A few years before this triumph, her team had undertaken gene-swap experiments, like those we encountered previously in relation to roots and leaves (Chapter Four), transferring moss genes *SMF1* or *SMF2* into *Arabidopsis* lines lacking copies of their own stomatal master genes (*SPCH*, *MUTE*, or *FAMA*).

They found that under guidance of *SMF1* but not *SMF2 Arabidopsis* mutants developed leaves with stomata.[34]

Although this proved to be an exciting step forward, it was not yet the whole story. Around the time we discovered that a single moss gene was the forerunner to one of the master genes discovered in *Arabidopsis*, we learned that distinguished professor Ralf Reski's research group at the University of Freiburg, Germany, had been carrying out similar experiments investigating the role of a gene called *SCRM* in stomatal formation in moss. Reski's obsession with moss has, amongst other things, seen him sequence its genome and start up a biotechnology company utilizing moss bioreactors for the production of pharmaceuticals. Reski's experiments had shown that *Physcomitrella* without *SCRM* developed normal-looking sporophytes that lacked stomata. We soon set up a Skype call and after summarizing the findings of our recently completed experiments to each other, decided to join forces and work together collaboratively. It turned out to be a wise strategic decision.

We knew that two genes (*SMF1* and *SCRM*) are essential for triggering stomatal development in moss and this raised the question: what mechanism links their action together? The explanation comes, once again, from what we know about how things operate in our model flowering plant, *Arabidopsis*. In *Arabidopsis*, *SPCH*, *MUTE*, and *FAMA* are unable to work alone. Instead, they work in partnership with *SCRM*. What actually happens is that the proteins coded for by *SPCH*, *MUTE*, or *FAMA* are produced at different times and places and chemically bind to the more widely expressed protein coded for by *SCRM*. On binding together, these proteins regulate the gene expression patterns guiding stomatal formation.[35] The question then becomes whether this partnership is ancient: does the moss SCRM protein interact with its SMF1 protein and lead to stomata? The short answer is yes, with the physical binding of the SMF1 and SCRM proteins activating stomatal development in moss in a manner analogous to how they work in flowering plants.[36] In fact, it turns out that nature had already done these sorts of genetic manipulations for us, and reached the same conclusion in a lineage of flowering plants—the seagrasses. Around 70–60 million years ago, the genome of the marine eelgrass, *Zostera marina*, evolutionarily lost its stomatal genes (*SPCH*, *MUTE*, *FAMA*, and *SCRM2*) and with them its stomata, structures that simply are not required for living a submerged lifestyle.[37]

These experiments with moss established that a pair of ancestral transcription factor genes (*SMF1* and *SCRM*) formed the core genetic machinery essential for stomatal development. Closely related genes are also present in a hornwort genome, the only other bryophyte lineage that makes stomata, although we have still to sort out the details of how they regulate the molecular pathways leading to stomata.[38] So this core genetic module for stomatal development existed in the last common ancestor of mosses, hornworts, and vascular plants, and this takes us closer to understanding the earliest events in stomatal evolution. We can also tentatively say something about how the genes enabling this early land-plant ancestor to build stomata might have worked.[39] The first sporophytes of tiny ancestral land plants that lived hundreds of millions of years ago probably operated with a pair of genes closely related to *SMF1* and *SCRM* that had multiple functions. The *SMF1*-like gene probably possessed *MUTE*- and *FAMA*-like activity as the minimum requirement for producing stomata in partnership with *SCRM*. In early land plants, *FAMA*-like activity may have specified guard cell identity and driven specific differentiation programmes leading to stomatal formation. *MUTE*-like activity then may have ensured that the guard cells matured into stomata. Here again is the extraordinary power of evo-devo in connecting molecular events in cells of living plants to ancient fossil structures found in long-extinct plants preserved in the rock record.

When it comes to stomatal evolution and the genetic trio of *SPCH, MUTE,* and *FAMA*, the story continued until, some 350 million years later on, evolutionary rewiring of closely related genes produced a fundamental innovation in grasses—a new type of physiologically improved 'super stomata'.[40] The stomata of most plants consist of a pair of kidney-shaped guard cells surrounding a central pore, but grasses evolved a second type composed of four cells—two dumb-bell shaped guard cells each flanked by linear subsidiary cells. Dumb-bell stomata are thought to be a customized version of their earlier-evolving kidney-shaped cousins. They tend to have faster responses to changing environmental conditions, such as a drying soil, thanks to the rapid transport of ions and metabolites between the guard cells and adjacent subsidiary cells[41] and the efficient mechanics of their operation.[42] This means they can snap shut faster to reduce water escaping in transpiration more effectively.

Advanced approaches recently led to the discovery that *MUTE* in a wheat-like grass species called *Brachypodium* regulates stomatal and subsidiary cell development, but that the protein coded for by the *MUTE* gene is larger than in *Arabidopsis* and can move between cells (hence it is known as 'mobile MUTE').[43] Mutant plants lacking mobile MUTE had two-celled stomata instead of the four-celled usual type. In comparison to the doctored mutants, the wild type of *Brachypodium* grew faster and had stomata that opened wider, and responded more rapidly to changing light intensity, than the mutants with 'normal' two-celled stomata. Most importantly, in growth experiments, two-celled mutants grew poorly, producing a third less biomass than normal plants. These exciting findings provide compelling evidence that speedy grass stomata help maximize the productivity of grasses under changing environmental conditions. An added bonus is that they support the speculative idea that dumb-bell stomata helped the remorselessly successful diversification and spread of grasses during the most recent era of Earth history, the Cenozoic (the past 65 million years), as the global climate dried and cooled.[44]

Genes for stomatal development represent only one part of the 'air-breathing' toolkit needed by land plants. Additional toolkits are required for taking strategic decisions about how to space them apart properly for optimal efficiency in regulating the exchange of gases across the leaf surface.[45] If you look at the arrangement of stomata on leaves under the microscope, it quickly becomes evident that they are beautifully ordered so that no single stoma is too close to its neighbour, each being separated by at least one epidermal cell. This spacing ensures the pore can open and close properly because, as the guard cells swell up, they elbow adjacent epidermal cells out of the way, gaining mechanical leverage in the process. Cells are, therefore, constantly in communication with each other, discussing who is going to do what by secreting small signalling compounds (called peptides) coded for by *EPF* genes. This molecular dialogue is a form of chemical Twitter, communicating in 'words' of 140 amino acids or less, and it operates before *SCPH/MUTE/FAMA* get to work, to hold them back and make decisions about when it is time to make a stoma.[46] Different versions of these signalling molecules are produced in differing amounts by neighbouring cells. Once synthesized, they diffuse outwards through cells and tissues, crowding on to special receptors poking out of the surfaces of neighbouring cells.[47] Decisions to make stomata are taken by the majority vote of the peptide molecules; when those synthesized by one set out-compete those of another on the receptor complex, it is time to advance toward making guard cells.[48]

Fossils of early land plants from Silurian and Devonian rocks tell us that this is an ancient mechanism because the stomata scattered over the surfaces of these simple vascular land-plant stems appear to obey the one-cell spacing rule, with little evidence of clustering. The components of the stomatal patterning toolkit used by these plants are inferred from those present in moss.[49] Moss has genes coding for the EPF peptides, receptor complexes, and so on, and if you switch off the EPF 'Twitter feed', it develops numerous clusters of stomata breaking the one-cell spacing rule. The moss stomatal patterning toolkit is simplified and wired up differently from the complex version found in flowering plants, but analysis of the evolutionary history of the genes shows the debt they owe to early land plants.

So far, in this chapter, we have seen that early land plants had at least two closely linked genetic toolkits for building stomata, the air-breathing hardware. One set decides when and where stomata are made (i.e., they specify patterning) and the other triggers the specialized development of guard cells (i.e., development). Stomata also require metabolic 'software' for making those 'water-for-carbon trader' decisions Berry mentioned. Flowering plants have fantastically complex control systems regulating stomatal opening and closing in response to things like light and drought. Decisions about when to open and close stomatal pores as the plant experiences constantly fluctuating environmental conditions are taken in response to chemical and environmental signals. Roots that detect drying soils, for example, release a mobile chemical messenger of drought in the form of the hormone abscisic acid,[50] usually abbreviated to ABA. The name is a historical legacy;[51] ABA was originally discovered to have a function in shedding (abscising) leaves. The mechanism by which this triggers stomatal closure is as follows: upon sensing ABA, guard cells extrude negatively charged ions, causing the membrane-spanning potassium channels to open and export potassium. This causes guard cells to shrink, closing the stomatal pore, and thereby helping to alleviate the shortage of water and turn off ABA signalling. It provides plants with a hormonal feedback system facilitating the conservation of water.[52]

Faced with the double penalty of shallow soils prone to drying out, and a lack of forest to cast relief from the Sun's glare, early vascular land plants would have needed an ability to fine-tune transpiration by sensing a drying soil. This raises an interesting fundamental question: how did these plants go about the crucial business of limiting water loss necessary to succeed on land? Historically, the first hint came from experimental work reported by botanists from Cornell University dating to the early 1970s. Working with a moss (rather than a flowering plant) called *Funaria hygrometrica*, they were able to show that it closed its stomata in response to the application of ABA.[53] Intrigued, we repeated this experiment nearly fifty years later to check if this really was the case. Sure enough, exposing the stomata of two moss species, including *Funaria* for good measure, to an increasing dose of ABA caused the pores to close progressively.[54] Taken at face value, these observations suggest that an extant early land-plant lineage (moss) had an operational core ABA signalling and response pathway for tuning water loss by transpiration. In fact, some of the genes involved are ancient and may have been originally used for desiccation tolerance before being co-opted for regulating stomatal functioning in early land plants.[55]

Encouraged by our simple experiments with moss stomata, we investigated the function of one of the key genes, called *OST1*, which sits in the middle of a metabolic pathway triggering stomatal closure by ABA in flowering plants.[56] Repeat the ABA experiments with moss lines lacking *OST1* and the stomatal closure response is lost. Introduce moss *OST1* into *Arabidopsis* mutants lacking their own copy and it restores the normal stomata closure responses to ABA. These genetic manipulations show that moss and vascular plants share a common functional component of the toolkit for responding to drought. A further detail is that *OST1* works by activating ion-pumping channels in guard cell walls necessary for stomatal closure. The gene coding for a protein that makes the membrane-spanning ion channel is called *SLAC1*. Fascinating work has shown that *OST1* can activate the SLAC1 ion channels responsible for the export of negatively charged ions to close stomata in mosses, lycophytes, and seed plants (groups with stomata), but not those from groups lacking stomata—liverworts and charophyte green algae.[57] We also know why this is the case. The protein coded for by *SLAC1* in liverworts and green algae has small differences in its amino acid sequence and this prevents it from being recognized and activated by the OST1 protein.

Others have added greatly to the overall picture by elegantly mapping the presence of genes and proteins related to the perception and metabolic signalling of ABA in numerous species of green plants, from chlorophyte algae through charophyte algae to land plants ranging from mosses and liverworts to a diversity of seed plants.[58] Such findings are often presented as coloured heat maps, formed by a matrix of rows and columns. Each horizontal row in the matrix represents a species belonging to a particular lineage and each vertical column represents a gene or its protein sequence involved in ABA metabolism. The blocks in the resulting matrix are shaded from pale yellow to deep red, depending on their similarity to the best-documented genes and proteins in *Arabidopsis*. The greater the similarity, the hotter the matrix glows.

At a glance, these sorts of analyses provide heat maps offering a snapshot of over a billion years of evolution and as you cast your eye over it, two things are immediately apparent. First off, you see an overall increase in the number of genes relating to ABA metabolism moving from algae to seed plants. Second, you see that the core ABA signalling pathway was established during the transition from an aquatic to a terrestrial environment over 400 million years during the origin of land plants, with the metabolism gaining in complexity as seed plant lineages radiated into drier late-Palaeozoic environments.[59] Experiments confirm the functioning of the pathways, as opposed to the presence/absence of genes, during the evolutionary transition to seed plants.[60] It is a compelling approach, combining molecular and physiological tools to sort out the evolutionary history of an essential adaptation of terrestrial plant life.

But we need to return to mosses, because their stomata may hold the key to understanding the function of stomata in the first complex vascular land plants. The question of function intrigued those Cornell botanists too, and they noted a vital clue: as the small moss capsules ripen through ageing, the responsiveness of their stomata to environmental stimuli declined until most of them appear to be 'locked' open as they developed thick rigid walls.[61] There can be few explanations for encouraging an uncontrolled loss of water from ageing capsules (sporophytes) and top of the list is reproduction. Second on the list is the idea that pores locked open might contribute to the transpirational pull of water and nutrients up into the capsule.[62] With the sporophyte structure attached to its gametophyte parent plant, it is not clear how important this proposed effect might be. Aiding reproduction is implicated because after spores develop inside the capsule, open stomata may facilitate drying of the capsule, a process causing its progressive contraction and gradual release and dispersal of spores.[63] Of course, if this idea is correct, the mutant mosses lacking stomata mentioned earlier could be used to test it. So we did, and this is exactly what we found. Mosses lacking stomata showed delayed maturation of their capsules and release of spores compared to normal mosses with stomata, as anticipated if stomata in this case functioned to promote water loss.[64]

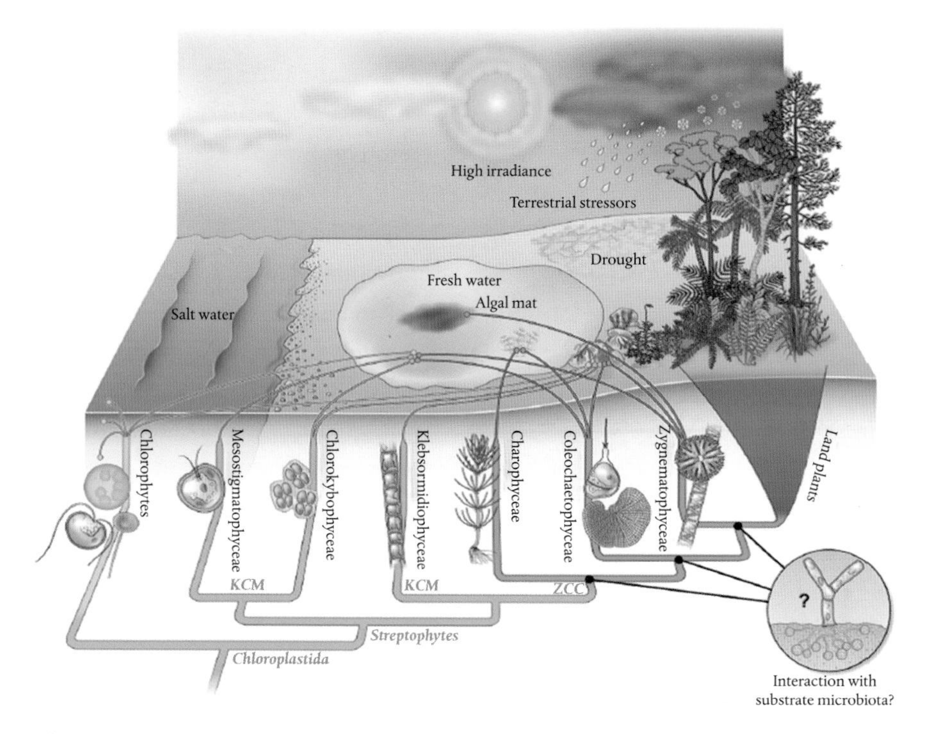


**Plate 1.** Colonization of terrestrial habitats by streptophyte green algae, the group that includes all land plants. ZCC abbreviates the higher branching grades (Zygnematophyceae, Coleochaetophyceae, and Charophyceae) and KCM abbreviates the lower branching or basal grades (Klebsormidiophyceae, Chlorokybophyceae, and Mesostigmatophyceae).

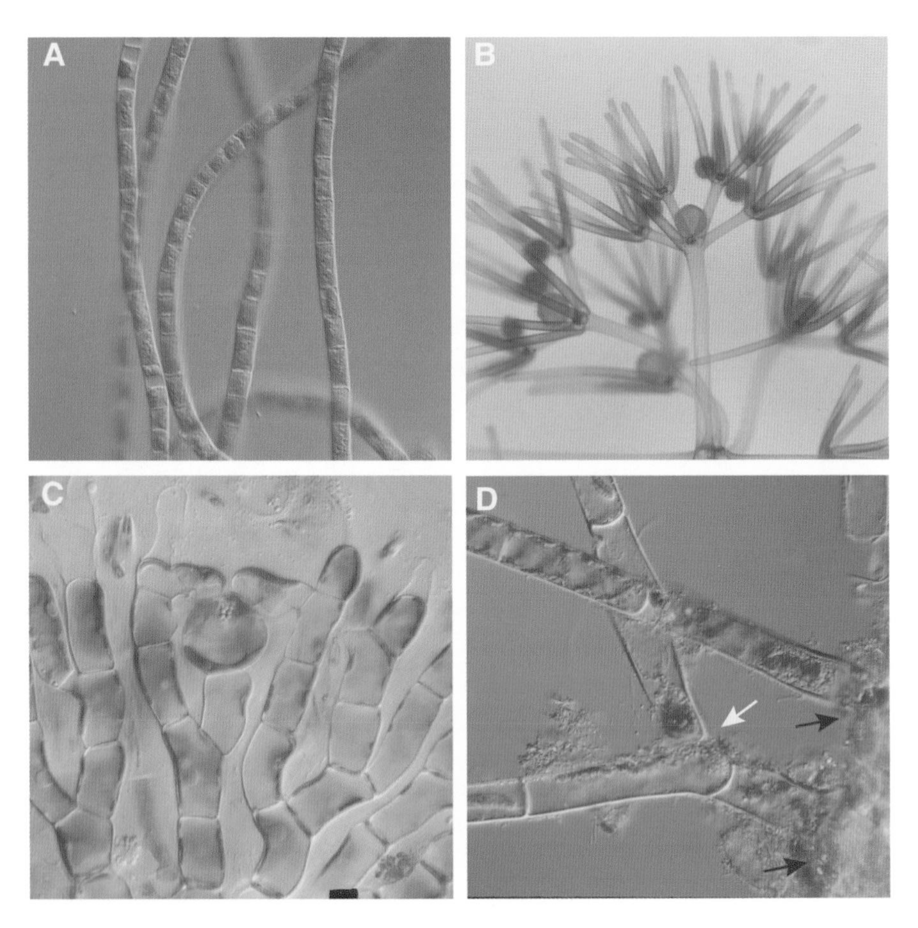


**Plate 2.** Present-day species of charophyte algae, the group which includes the freshwater ancestors of land plants. (A) *Klebsormidium nitens* (Klebsormidiales), (B) the stonewort *Nitella hyalina* (Charales), (C) *Coleochaete pulvinata* (Coleochaetales), and (D) *Spirogyra* (spiral chloroplast) and *Mougeotia* (flat chloroplast) (Zygnematales). Note basal branch in *Mougeotia* (white arrow) and holdfasts (black arrows).

**Plate 3.** The diversity of plants. (A) An assemblage of *Phaeoceros* (a hornwort; white arrow), *Fossombronia* (a leafy liverwort; red arrow), and interspersed mosses. The arrows point to the sporophytes. (B) *Lycopodium digitatum*, a lycopod, showing spore-bearing cones. (C) A tree fern (*Cyathea horrida*), (D) the cycad *Cycas revoluta*, a widely cultivated cycad sometimes called 'Sago Palm', (E) *Nymphaea* hybrid, a water lily, a representative of one of the basal branches of flowering plants, and (F) *Ampelopsis* sp., grape family (Vitaceae) being pollinated by a wasp. *Ampelopsis* represents the eudicots.

**Plate 4.** Coastal redwoods, the tallest trees on Earth. This titan lives in Prairie Creek Redwoods State Park, California, USA and is probably over 1500 years old. Photo composed of a mosaic of 84 images.

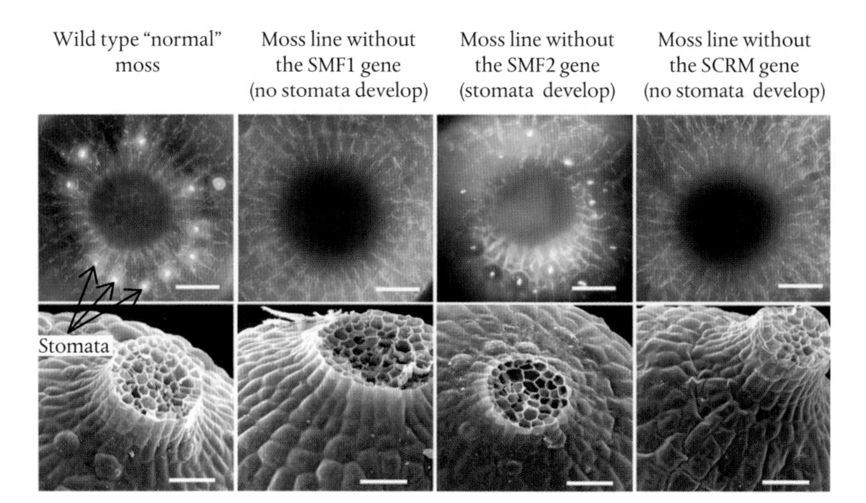


**Plate 5.** Normal moss develops stomata on its sporophytes, whereas lines lacking *SMF1* and *SCRM* genes develop sporophytes lacking stomata. In contrast, those with the *SMF2* gene develop normal stomata. The top line of images were taken with an epifluorescence microscope which causes the stomata to glow. The lower set are photos taken using a scanning electron microscope. Scale bar in all images = 50 μm.

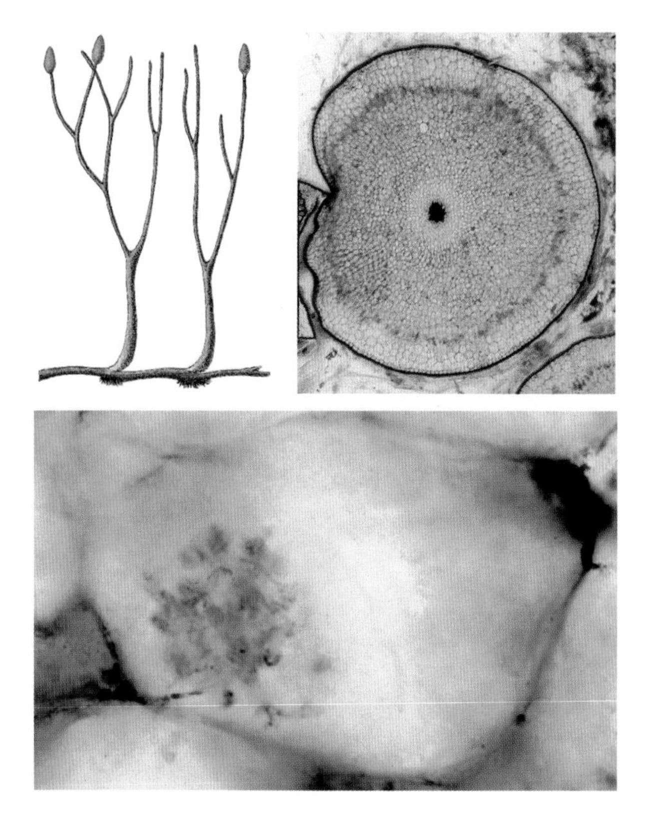

**Plate 6.** Reconstruction of the early vascular land plant *Aglaophyton* with images of a cross section of a rhizome and fossilized fungal structures resembling arbuscules. Top right photo (cross section): ×15 magnification; bottom photo: × 600 magnification.

**Plate 7.** Exhibit of sedimentary tree stumps outside the Gilboa Museum.

**Plate 8.** Reconstruction of the 385-million-year old complex forest at Gilboa, one of the first forests on Earth.

**Plate 9.** Mobile truck-mounted rig for drilling rock cores at a quarry near Cairo, New York State.

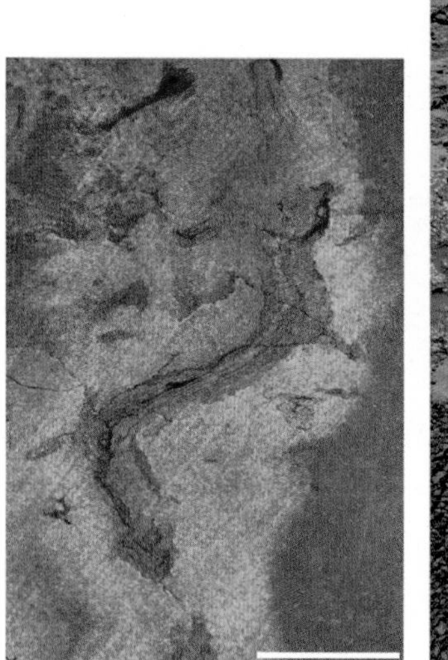

**Plate 10.** Rock cores of fossil soils drilled from a 385-million-year-old forest floor. Smaller image shows small tree (archaeopteridalean-type) root, termed a rhizolith, preserved as clay cast with central carbonaceous strand, surrounded by a 'drab-halo'. Scale bar = 1 cm.

Applied to understanding the functioning of Silurian–Devonian vascular land plants, these ideas suggest that the basic genetic toolkits for building and operating stomata provided them with a means of both staying hydrated and generating a transpiration stream for accessing soil water and dissolved nutrients when it was safe to do so. Later, during the reproductive phase, the stomata might have switched roles and promoted transpiration to cause release of spores from sporangia, so facilitating colonization by the next generation. This emerging picture of stomatal function in early land-plant sporophytes ultimately links stomatal action to reproductive success.

The story from the DNA records of modern plants is a telling one: the origin of stomata in the last common ancestor of mosses, hornworts, and vascular plants ~500 million years ago coincided with the evolution of a core genetic module enabling ABA metabolism signalling that links limited water supply to limiting water loss via stomata. Another way to think about the significance of this new view of the world is like this. Next time you take a walk through the countryside on summer's day, it may be worth reflecting on the long and eventful history of stomata in the leafy canopies above your head, or the grass blades brushing against your knees. They are sensing the environment in the soil, the atmosphere, and in their surrounding cells and processing the information to make decisions about opening and closing. Collectively, millions of them are cooperating to maximize the beneficial uptake of carbon dioxide to grow the plant and minimize the loss of precious water supplies. Making these decisions are genes derived from an ancient core genetic toolkit that proved of fundamental importance to the earliest land plants nearly half a billion years ago.

Today, tiny, exquisitely sophisticated stomata densely stud the leaves of virtually every species of plant on the planet. Only a handful of specialized amphibious species can live without stomata. Instead of capturing carbon dioxide from the air, these plants draw it up through their roots in dissolved forms from the sediments.[65] The collective planetary gulp of carbon dioxide by the global population of stomata on the leaves of land floras is an impressive ~430 billion tonnes each year.[66] About half of this is converted into new plant growth; the other half is released back into the atmosphere by respiration by plants and microbes. The enormous uptake and release of carbon dioxide by land plants creates a seasonal cycle in its atmospheric concentration, detectable by air measurements made across the world, which really reflect the 'breathing' of land floras. By permitting land floras to safely refuel with carbon dioxide and grow, stomata help sustain a huge diversity of animals, from elephants to earwigs. They play an essential role in feeding the voracious appetite of herbivores across the globe. Unless you live exclusively on a diet of algae-grazing fish and squid, the same holds for humans too—most of us are composed of carbon atoms that have passed through stomata pores.

John Updike, in *The Centaur* (1962), takes up this theme, writing that green plants 'take in moisture and the carbon dioxide we breathe out and the energy of the sunlight, and they produce sugar and oxygen, and then we eat the plants and get the sugar back and that's the way the world goes around'. To which we can now add that the carbon atoms in those carbon dioxide molecules have travelled through microscopic stomatal pores before the biochemical machinery processes it via by photosynthesis and makes it into sugar. Updike later writes poetically that 'ultimately the energy for photosynthesis comes from the atomic energy of the sun. Every time we think, move or breathe, we're using up a bit of golden sunshine'. Human society obtains energy by releasing the energy of golden sunshine trapped in fossil fuels, like coal. Coal is composed of concentrated carbon atoms that travelled through prehistoric stomata and were converted into biomass by prehistoric photosynthesis; for as everyone knows, coal is the fragmentary remains of ancient tropical swamp rainforests.

Unlike the rainforests of the primeval swamplands, flowering plants dominate modern tropical rainforests and, luxuriating in year-round warmth, they are the most productive forests on Earth. But the remarkable productivity of these rainforests is not simply a function of the warm climate. It is made possible because the evolution of stomata in flowering plants progressed hand-in-hand with evolution of the plumbing systems (veins) that distribute water to photosynthesizing leaf cells. Anatomically, the valves and tubes that make leaves work are arranged so that the veins supplying water to the leaf deliver it close to the sites of evaporation in photosynthesizing tissues, near to where stomata are located. The arrangement helps prevent leaves from dehydrating when stomatal pores open to allow the inward diffusion of carbon dioxide. As a rule of thumb, we can say that having plumbing systems better able to supply water to leaves is a good way of supporting productive plants. It is the core principle Joe Berry was alluding to with his Wall Street trader analogy, with leaves better able to trade more soil-water for atmospheric carbon dioxide. But the hidden cost to building leaves with better hydraulic properties is that water-conducting tissues are expensive to construct, in terms of carbon.[67]

Why has evolution driven the selection of leaves with a high capacity for transpiration given a high carbon penalty for their construction? The answer seems to be that it is partly the evolutionary legacy of dwindling atmospheric carbon dioxide levels since the age of the dinosaurs, early in the Cretaceous.[68] Slowly falling carbon dioxide levels may have forced flowering plants to evolve leaves with larger numbers of miniaturized stomata to maximize their uptake of carbon dioxide for growth.[69] This, in turn, drove the miniaturization of leaf-vein systems with the appearance of fine veins forming dense meshes to better supply water to leaf cells doing the photosynthesis.[70] Only flowering plants evolved the novel developmental mechanism enabling this to happen and it gave them a much greater vein density per unit area of leaf.[71] Fossil leaves document the pattern of increased investment in water supply tubes since the Cretaceous that helped ensure flowering plants could grow faster.[72] According to the fossils, no other group of land plants—lycophytes, ferns, or gymnosperms—followed this evolutionary trajectory. Without it, leaves are not sufficiently well irrigated to safely open their stomata to absorb carbon dioxide to increase photosynthesis productivity. In effect, falling carbon dioxide levels, over millions of years, locked plants into a 'hydraulic arms race' between species. Armed with an ability to miniaturize stomata and water-supply vein networks, flowering plants had the edge and began to eclipse ferns and conifers in the world's tropical zones.

The leaves in modern tropical rainforests are a legacy of these past events and today these forests are the transpiration engines of the world, air-conditioning the planet by extracting water from deep in the soil profile and pumping it back to the atmosphere. In the Amazon, up to half the rainfall is recycled in this way each year and it makes the region around 5°C cooler than it would otherwise be, and substantially wetter.[73] By helping to generate wetter tropical climates and complex habitats, tropical forests foster their own existence and a rich diversity of life. What might happen as we pursue a programme of deforestation, degrading the land surface until it is covered by vegetation like ferns, lacking the high transpiration rates of tropical trees? In a world without rainforests showering the atmosphere with transpiration, tropical climates, especially those in South America, would be hotter, drier, and more seasonal.[74] Current and future deforestation of the Amazon will affect the water cycle. Cut down the forests and you halt the delivery of moisture to the air above, inhibit water recycling, and reduce rainfall. Some projections of forest loss by 2050 indicate it will cut rainfall by 20% in the dry season across the Amazon basin[75] and exterminate species. At its simplest, the message here is this: no rainforests → no rain → extinction.

The evolutionary dance between valves, tubes, and the carbon dioxide content of the atmosphere continues, with global consequences for Earth's future climate. For stomata will have a say in how the planet's climate system responds to the increasing atmospheric carbon dioxide concentration resulting from our ongoing combustion of fossil fuels. Decisive and meticulous experiments undertaken by the English plant physiologist Oscar Victor Sayer Heath (1903–1997), soon after the Second World War,[76] established that an atmosphere enriched in carbon dioxide causes stomata to close partially.[77] Heath's seminal experiments, undertaken on the stomatal behaviour of wheat, are fast becoming a new reality as we realize they hold for the majority of land-plant species. Already, the stomata of northern hemisphere forests are closing in the manner Heath anticipated, as atmospheric carbon dioxide levels climb and re-tune the physiology of trees to affect the wider planetary environment.[78] The wider environment here is the global climate system and the significance of Heath's discoveries for climate change are explained by the distinguished stomatal physiologist Terry Mansfield, as follows: 'This alters the rate of transfer of water from the soil to the atmosphere, and it also affects the surface–atmosphere exchange of heat and contributes to global warming. Thus the ability of stomata to sense and respond to $CO_2$ [carbon dioxide] in the atmosphere, once thought to be an obscure topic only of academic interest to Heath and a few other scientists, has become a major factor in our understanding of the forces that are driving climate change.'[79]

As Mansfield explains, when stomata close, the transpiration stream that seeds rainfall and air-conditions the planet begins to shut down. Changes in atmospheric water vapour, clouds, and the exchange of energy between the land surface and the atmosphere follow. It is obviously challenging to assess the extent of carbon dioxide-related stomatal feedbacks on future climate change. Simulations of land carbon and water cycling by forests, and linkages within the global climate system, are uncertain on the detail.[80] Nevertheless, in general they confirm that when the effects of elevated carbon dioxide on stomata are included, they modify both the climate and the hydrological cycle.[81] Simulations of a carbon dioxide-rich atmosphere show that its simple effect of causing stomatal pores to open less widely contributes some 30% additional warming across large areas of the boreal forests and the tropical rainforests (**Figure 16**). This additional warming is *over and above* that caused by the enhanced greenhouse effect resulting from the carbon dioxide-rich atmosphere (i.e., by trapping long-wave radiation).

Half a century on from Heath's important discovery concerning how carbon dioxide affects stomatal behaviour, we are discovering key genes involved in pathways mediating the perception of carbon dioxide and how it controls their development. Surprisingly, the important genes involved in perception of carbon dioxide in plants are remarkably similar to those possessed by humans and insects, even though the evolutionary pathways of humans and plants diverged over a billion years ago.[82] Why mammals evolved a mechanism to sense carbon dioxide is unclear. Was it to alert them to rotting food, which releases carbon dioxide? Insects, like fruit flies, moths, and mosquitoes, sense carbon dioxide to find food items like decaying fruits, flowers, and people. The genes responsible function interchangeably between these groups of organisms and code for an enzyme called carbonic anhydrase. Carbonic anhydrase combines carbon dioxide and water and converts them into a simple acid (carbonic acid) plus bicarbonate

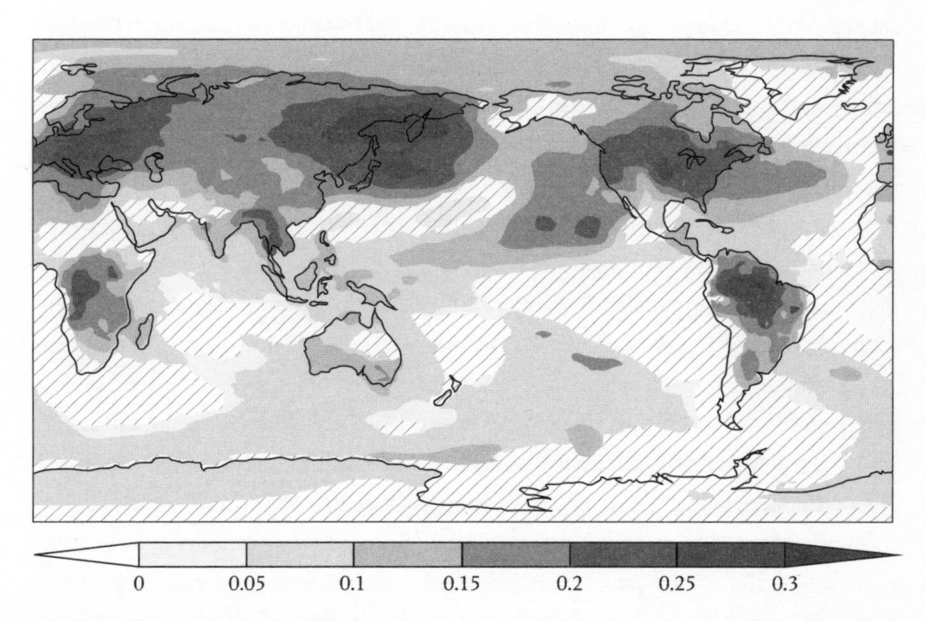


**Figure 16** Fraction of total surface warming caused by the physiological effects of carbon dioxide on stomatal behaviour.

ions. The acidic signal interacts with the sour-cell taste-bud receptors on our tongues. In leaves, the bicarbonate signal activates molecular pumps in the walls of the guard cells that trigger the pore to close. If you transfer the gene for mammalian carbonic anhydrase into *Arabidopsis* plants lacking their own version, it restores the responsiveness of stomata to carbon dioxide.[83]

Plants are also adapting to rising atmospheric carbon dioxide levels in another important way: by reducing the number of stomata developing on their leaves. Evidence from historical collections of leaves of trees from southern England suggests trees noticed the rise in atmospheric carbon dioxide levels and took action on it in this way,[84] before we had started measuring the concentrations for ourselves.[85] Steps towards solving the mystery of how plants reduce the numbers of stomata came thirty years after that evidence came to light, with the discovery that leaves sensing rising atmospheric carbon dioxide levels activate their 'Twitter feed' to produce more of a particular peptide, called EPF2.[86] This blocks the formation of stomata in the epidermis of leaves at elevated carbon dioxide concentrations. Other proteins known as proteases activate the EPF2 peptide. One in particular (coded for a gene called *CRSP*, Carbon dioxide Response Secreted Protease) is crucial for activating EPF2 and is responsive to atmospheric carbon dioxide levels. This part of the newly discovered pathway acts as a carbon dioxide-regulated volume control on gene expression, offering a sensitive means of balancing stomatal development as the atmospheric carbon dioxide concentration changes. You can imagine that such a 'sensing and response' mechanism involving two genes (*CRSP* and *EPF2*) may offer scope to engineer crop varieties for improved performance with a changing global climate.

An early warning sign that rising atmospheric carbon dioxide levels might be closing stomata, or causing leaves to be made with fewer of them, is the recent increased delivery of freshwater from the land to the oceans. Roots take up less water from the soil if forest canopies are constructed of less porous leaves, all else being equal, and this means more water drains from soils into rivers and ultimately into coastal waters. British scientists have reported that the terrestrial biosphere is already feeling the effects of our carbon dioxide emissions and responding in this way.[87] According to their analyses, a trend of increasing amounts of freshwater flowing off the continents to the oceans began around the beginning of the twentieth century, when carbon dioxide levels started to climb more steeply. Climate and land use over the same period are ruled out as possible causes, meaning that it is the direct effect of rising carbon dioxide levels, linked to the stomatal suppression of evaporation from global forests.

Back in 1976, the UK suffered a severe heatwave and widespread drought killing trees and causing crop failure. That same year the distinguished meteorologist John Monteith (1929–2012)[88] closed a prescient meeting at the Royal Society, London, on plants and the water cycle by quoting the Victorian poet Alfred, Lord Tennyson's lines written at the time of the potato famine in Ireland:

> Science moves but slowly, slowly, creeping on from point to point
> Slowly comes a hungry people, as a lion creeping nigher,
> Glares at one that nods and winks behind a slowly dying fire.

Crops and crop production were very much on Monteith's mind, and with intellectual poise he suggested Tennyson's lines had modern relevance. He asked 'whether we are the people nodding and winking behind sophisticated research projects while hunger and malnutrition remain an immense global problem'. Urgently improving our ability to feed a growing human population that may reach 11 billion by 2100 is very much a modern research priority. Stomata are now prime targets for boosting yields of major crops by making them more efficient in using water in a warmer, drier future climate. Continued progress in this area depends on better understanding the complex development and physiology of stomata, and the roles they play in allowing plants and global ecosystems to manage their watery economics.[89]

Meanwhile, continuing our progress in understanding how global ecosystems 'greened' the continents means we need to think more broadly. We need to move beyond thinking in terms of accumulating a succession of evolutionary adaptations to build successful land plants. Plants also succeeded on land by forging alliances with another group of organisms—soil microbes. That story, which set in motion a chain of events that was to fundamentally transform global ecology and the chemistry of oceans and atmosphere, unfolds in Chapter Six.

# 6 ANCESTRAL ALLIANCES

'Sam Gamgee planted saplings in all the places where specially beautiful or beloved trees had been destroyed, and he put a grain of precious dust from Galadriel in the soil with the root of each.... Spring surpassed his wildest hopes. His trees began to sprout and grow, as if time was in a hurry and wished to make one year do for twenty.'

J.R.R. Tolkien, *Lord of the Rings* (*The Return of the King*, 1955)

Rhynie is a small, unremarkable village in a fertile valley of the Old Red Sandstone area in Aberdeenshire, Scotland. Amongst palaeontologists, though, Rhynie is internationally renowned for exquisitely preserving the fossilized anatomy of a terrestrial ecosystem of the early Devonian Period. Beneath the green fields and mantle of soil lies a snapshot, frozen in time, of terrestrial life from 400 million years ago that includes primitive land plants, fungi, algae, lichens, tiny invertebrate animals, and bacteria. This extraordinary site was an ancient hot spring environment, analogous to that found in Yellowstone National Park, and its unusual setting explains the exceptional preservation quality of the fossils. As hot, silica-rich hydrothermal fluids bubbled up from below, they inundated the cells and tissues of the primitive flora and fauna surrounding the warm pools, preserving astounding microscopic detail. The resulting fossils are contained in a hard, siliceous fine-grained form of rock known as chert, and specimens must be prepared by cementing pieces of the rock matrix to microscope slides and then grinding and polishing them to produce near-translucent 'thin sections'. Laborious though such efforts may be, they can deliver spectacular results. Land plants at that time reproduced by the fertilization of female sex organs with sperm, and fossils prepared in this way record the young sperm developing inside the male sex organs. On release, the motile sperm would have swum through the watery environment and fertilized the egg cells, just as happens today in the bryophytes, lycophytes, and ferns.

The Scottish doctor and proficient part-time geologist William Mackie (1856–1932) made the sensational discovery of the Rhynie lagerstätte in 1912, quite by chance.[1] 'Lagerstätte' is the name given to an exceptionally preserved fossil assemblage, derived from the German *lager* and *stätte*; literally meaning 'place of storage'.[2] One version of the story is that, after a strenuous morning of geological mapping, Mackie sat down on a dry stone wall to eat his lunch and reached for a nearby loose chunk of chert. On inspecting the rock in his grasp, possibly between mouthfuls of sandwich, Mackie was stunned to discover it was packed with fossilized plant stems. Alternatively, it may be that it was only after he returned to the lab and made thin sections of the blocks of chert found in walls, and littered around in fields, that the spectacular nature of his discovery started to become clear. The fossil collector, and member of the Geological Survey, Mr Tait was enlisted to undertake later exploration of the site, funded by the Royal Society, London, and the British Association.[3] This involved digging a trench through the top soil (**Figure 17**) and excavating blocks of chert suitable for cutting petrographic thin sections to reveal the spectacular riches of fossilized early Devonian life on land.[4]

The formidable and accomplished palaeobotanical duo of Robert Kidston (1852–1924) and his younger friend and colleague William Lang (1874–1960) examined the thin sections. In the resulting series of papers, published in the *Transactions of the Royal Society of Edinburgh*, Kidston and Lang documented in extraordinary detail the external appearance and anatomy of the small-stature early land plants, and other organisms, found in the silicified deposits.[5] These classic works included beautifully illustrated accounts of primitive vascular plants with upright leafless stems less than 20 cm tall and horizontal rhizomes adorned with rhizoids (**Figure 18**) and are now regarded as 'the most important contributions ever made on the knowledge of the plants of the Devonian'.[6] Discussions initiated by Kidston and Lang concerning the evolution of roots, shoots, and leaves, and the nature of plant life cycles, later proved foundational to the field of 'evo-devo' from the world of molecular biology, as we have seen.

In the final paper of the series, Kidston and Lang dealt extensively with the fungi associated with the plant remains. There they wrote about how fungi appeared localized inside the tissues of the rhizomes and lower stems of the fossil plant

**Figure 17** The village of Rhynie, Aberdeenshire, Scotland. Photograph below shows exposure of the fossiliferous chert beds beneath the mantle of soil.

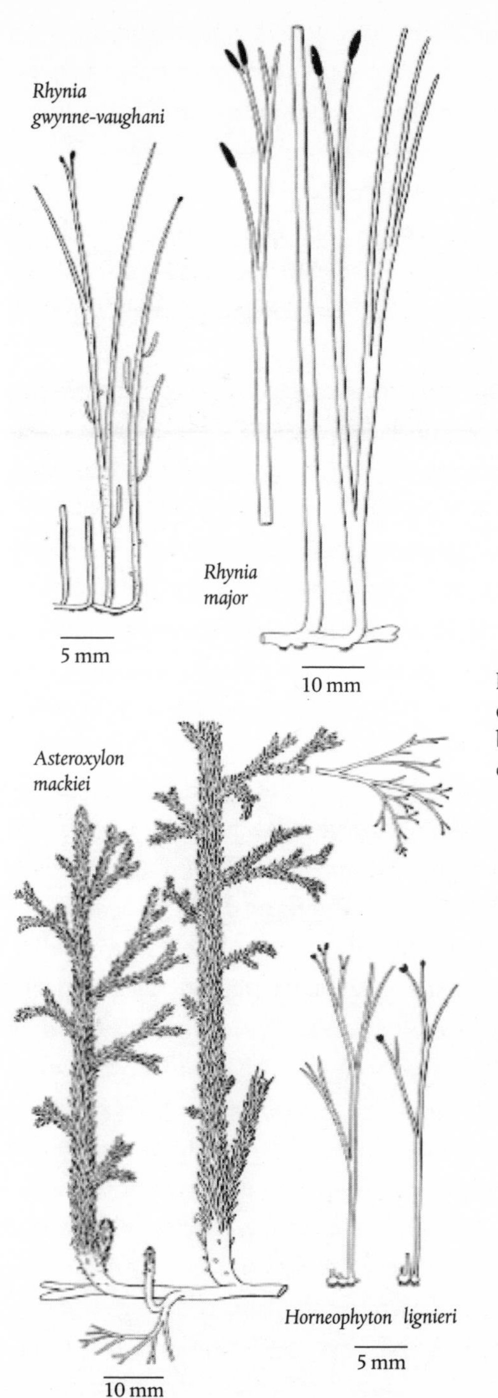


**Figure 18** Original reconstructions of early Devonian fossil vascular plants by Kidston and Lang from the Rhynie chert.

*Asteroxylon*, writing: 'the localization of the fungus to the inner cortex, leaving both the outer cortex and the phloem free, is in support of the view it was mycorrhizal'. They went on to cautiously draw the conclusion that, 'although the evidence of a mycorrhizal relation is considerably stronger [than for *Rhynia*], it does not amount to proof'. Their reluctance to draw definitive conclusions hinged on the difficulty of distinguishing between saprotrophic fungi that invaded the plant material after its death, and symbiotic fungi that interacted beneficially with living plant tissues. Indeed, the majority of the fungi described based on fossil spores/cysts and filaments were saprotrophs.

Clinching evidence followed a century later with the discovery that fossilized stems of another primitive plant from Rhynie, called *Aglaophyton majus*,[7] contained beautifully preserved fungal arbuscule structures (from the Latin *arbusculum*, or 'small tree') (**Plate 6**).[8] Fossils of these highly branched fungal structures were present within the cells of *Aglaophyton* rhizomes and prostrate stems, which the authors described as being 'morphologically identical to those of living arbuscular mycorrhiza in consisting of a basal trunk and repeatedly branched bush-like tuft within the plant cell'.[9] Later research discovered arbuscules in the gametophyte phase of the plant's life cycle as well.[10] Arbuscules are the characteristic structures of symbiotic arbuscular mycorrhizal fungi formed within the living cells of modern vascular plant roots by, whose importance for the symbiosis will become clear shortly. Other studies of Rhynie fossils have since discovered fungal spores with characteristic shapes and distinctive cell-wall structures indicative of several different groups of these fungi, leaving no doubt that arbuscular mycorrhizal fungi had evolved by this time.[11]

Mycological discoveries from the Rhynie chert set the stage for the later publication of one of evolutionary biology's most provocative ideas, put forward by Polish-born Kris Pirozynski[12] and his colleague, the Canadian David Malloch (**Figure 19**). In their classic 1975 paper[13] entitled 'The origin of land plants: a matter of mycotropism', Pirozynski and Malloch argued that a beneficial partnership between mycorrhizal fungi and land plants catalysed the origin of our terrestrial floras. If you met either of our two mycological mavericks today, neither would strike you as a revolutionary. Since his retirement, Pirozynski has turned his back on mycological matters and become preoccupied with stamp collecting. Malloch, on the other hand, continues to teach mycology courses to enthusiasts at weekends and vacations at the University of Toronto, where he is an Emeritus Professor.

**Figure 19** Mycological revolutionaries, Kris Pirozynski (left) and David Malloch (right). Forty years ago, they proposed that soil fungi facilitated plants' transition to land.

In 1975, they put the argument like this: 'Terrestrial plants are the product of an ancient and continuing symbiosis of a semi-aquatic ancestral green alga and an aquatic fungus'. According to their theory, 'the Silurian–Devonian "explosive" colonization of land, and indeed the very evolution of plants, was possible only through mutualistic partnerships—partnerships that were equipped to cope with the problems of desiccation and starvation associated with a terrestrial existence'. The following statement staked out their position: 'Our claim is that land plants never had any independence [from fungi], for if they had, they could never have colonized the land'. And they only blotted their copybook by backing the wrong horse in claiming that a fungal group called oomycetes was involved. Oomycetes were classified as fungi at that time, but we now know they form a distinct lineage of fungus-like filamentous eukaryotic microorganisms, not symbiotic fungi. Nevertheless, as we shall see, the notion of fungal symbioses with early land plants has gathered more and more supporting evidence ever since their pioneering paper four decades ago.

We can begin to understand quite what a brave step it was to introduce the concept of symbiosis as a means of explaining the early appearance of land plants by considering how evolution was understood at the time. Darwin's mechanism of evolution works through the competitive process of what the English philosopher Herbert Spencer (1820–1903) dubbed 'the survival of the fittest', but said nothing about the 'arrival of the fittest'. Symbiosis, the co-operation between distantly related organisms, rather than competition within or among closely related species, is a concept that requires a radical shift in thinking, and weighs in on the arrival-of-the-fittest question. Frank Ryan's book popularizing symbiosis, *Darwin's Blind Spot*, documents the gloves-off heated battle that took place in scientific circles before Darwinians accepted symbiosis as a valid mechanism of evolution.[14] The point is that it challenges the idea of natural selection by competition as the only mechanism of evolution, and sees both partners in the symbiosis increase their fitness, rather than the expected outcome with one winner and one loser. This central argument of a zero-sum game also helped explain the speed of evolutionary change. In the traditional view, natural selection operating on a mutation-by-mutation basis was simply too slow to account for the 'sudden' appearance and diversification of plant life on land. The conquest of a radically different and hostile terrestrial environment from the watery milieu of freshwater algae was possible only (in their eyes) through a major evolutionary breakthrough, on a 'scale equal to or surpassing that responsible for the Cambrian "explosion" of animal diversity'.

Arbuscular mycorrhizal fungi, whose remains sit entombed within the fossil plants of Rhynie, now form one of the most widespread symbioses on Earth. Found in the roots of around 80% of all plant species, these fungi forge a symbiotic alliance beneficial to both partners.[15] In modern floras, host plants receive inorganic mineral nutrients scavenged by the fungi from the soil. In return, these plants provide the fungi with sugars and lipids (fats) produced by photosynthesis to enable them to grow, forage for nutrients, colonize other roots, and undergo asexual reproduction by spores. The exchange of materials—sugars in one direction and inorganic nutrients in the other—occurs across the arbuscules inside root cells. The tree-like shapes of these sophisticated symbiotic interfaces generate a large surface area for the efficient two-way exchange of materials to take place. Note that arbuscular mycorrhizal fungi are obligate symbionts, meaning they are dependent on plants for providing them with energy to proliferate and reproduce. Unlike their saprotrophic fungal cousins, they lack the capacity to obtain energy by decomposing organic matter.[16] Not all plants, on the other hand, are dependent on mycorrhizal fungi.

High-precision radiometric measurements on the rocks and minerals date the Rhynie site at 407 million years old.[17] This date broadly accords with the estimated age range for the probable origin of arbuscular mycorrhizal fungi based on

## OVERTHROWING SCIENTIFIC ORTHODOXY

The hypothesis that plants might form beneficial partnerships with soil-borne fungi reaches back to the nineteenth century, and had a culinary impetus. The German botanist Albert Bernhard Frank (1839–1900) deduced the essential details in the late 1880s when His Excellency, the Minister of Agriculture, Domains, and Forestry for the Kingdom of Prussia understandably commissioned him to investigate the possibility of cultivating truffles. Truffles are really the subterranean reproductive structures of fungi that grow attached to the roots of oaks and other tree species. The invitation changed the course of Frank's scientific career and led him on a decade-long quest to understand plant–fungal collaborations. With simple but elegant and pioneering experiments, combined with careful observations of the roots of trees, he challenged the then current orthodox opinion that all fungi were harmful (i.e., pathogenic) to plants. In early experiments, he collected seed from pine forests and sowed one set in soil in its natural state and another in sterilized soil. Those seedlings germinating in natural soil formed fungal partnerships and developed into stronger healthier saplings than those that grew on the sterile soil lacking fungi. Tolkien, who was 'much in love with plants, and above all trees', made detailed descriptions of them in his writing and appeared to understand the benefits of their symbiotic partnerships with soil fungi. He saw the need to add '*a grain of precious dust from Galadriel in the soil with the root of each* [tree]', noting how it stimulated sapling growth.

In proposing the revolutionary theory of tree nutrition involving a symbiosis between roots and fungi bound into a single organ—a mycorrhiza—Frank faced formidable opposition from incredulous colleagues. It took 50 years to settle the matter decisively but he never did crack the puzzle of cultivating truffles. Cultivation of Périgord or black truffle (*Tuber melanosporum*) and summer or burgundy truffle (*Tuber aestivum*) is now possible in Europe mainly through methods that mirror Tolkien's idea—spore inoculation of the soil into which the host tree is planted.[18]

molecular clocks.[19] Rocks and molecular clocks, then, point to the great antiquity of associations between early land plants and arbuscular mycorrhizal fungi. Yet the fossils from Rhynie, those compelling messengers of past events, remain silent when it comes to telling us about the nature of the interactions between these two groups of organisms,[20] leaving the presumed status of the symbiosis uncertain.

Not surprisingly, Pirozynski and Malloch's paper met with a sceptical reaction. Reflecting on it years later, Malloch remarked, 'most of my botany department colleagues thought it was pure fantasy'.[21] You can only suspect that, at the time, they were not alone. Nonetheless, Lynn Margulis, the prominent evolutionary biologist and great champion of symbiosis as a powerful creative force driving the evolution of life on Earth, promoted their extraordinary suggestion in her book *Symbiotic Planet*.[22] There she asserts that 'the great expansion on land up and out of both sea and freshwater was grounded in the intimacy of plant and fungus, and still is'. The evidence for this was little more than that which Pirozynski and Malloch relied upon in 1975—mainly fossils from Rhynie. Contrast this to the more cautious reading of the same fossil evidence many years earlier by Kidston and Lang, who wrote 'The appearances [of fungi] in the Rhynie plants do not afford such evidence [of symbiosis], although they leave open the possibility that a mycorrhiza may have been present in the actively living plant'.

So, with what might be regarded as a combination of brilliant insight and good luck, the duo offered the scientific community a major hypothesis that has hardened over the years into the current paradigm: mycorrhizal associations catalysed the origin and evolution of our land floras. Implicit in the argument is the idea that such 'mycorrhizal associations' were mutualistic, with both partners benefiting, like their counterparts in modern plants. Yet, surprisingly, this critical assumption has remained firmly in the realms of speculation for decades. Without definitive experimental evidence of how the partnership concealed in the fossil record might have functioned, it is difficult to say what, if any, benefits arose from the alliance. Scientific attention had instead focused exclusively on the details of the cellular interface between fungi and the cells of host plants such as liverworts, lycophytes, and ferns,[23] without knowing whether the exchange of materials between organisms essential to understanding potential symbiotic interactions actually occurred.

A few years ago, we teamed up with my mycological colleague, Jonathan Leake, and other collaborators, in an attempt to discover the nature of the alliance between early-evolving land-plant lineages and their fungal partners. The idea was to undertake experiments that blended mycology, ecology, and palaeontology. In doing so, we hoped to open a window for the first time on how this ancient partnership might have functioned during the greening of the Earth hundreds of millions of years ago. We focused on liverworts, as one of the oldest surviving lineages of land plants, and representatives of early colonists forming associations with mycorrhizal fungi.[24] The rhizoids of liverworts provide the primary pathway of mycorrhizal fungal colonization, and this also seems to have been the case for some of the early land plants at Rhynie.[25] The investigations began with a liverwort species by the name of *Marchantia paleacea,* which originated from soils beneath the temperate cloud forests of Veracruz, Mexico, and the cells and tissues of which are colonized heavily by arbuscular mycorrhizal fungi.[26] Our small breakthrough came when we discovered how to manipulate the fungal colonization of young plants nurtured from tiny vegetative propagules (gemmae) sitting in splash cups on the surface of the thallus. Once displaced from splash cups, gemmae develop into small plants, known rather charmingly as gemmalings.[27] These can be cultivated either in sterilized soil or adjacent to mature mycorrhizal adult plants to transfer fungal colonization. By means of this simple procedure, we could establish gemmalings of relatively uniform size and shape either with or without their symbiotic fungal partners, and this set the stage for some much-needed experiments. Before this, the roadblock to investigating the Pirozynski and Malloch hypothesis had always been that mycorrhizal fungi resisted all attempts at being grown and manipulated under laboratory conditions. The solution to the problem was to turn it around and manipulate the fungal colonization of the host plants.

Evidence gathered from experiments conducted with the gemmalings soon made clear what the fossil record could not tell us: liverworts gained substantial benefits from entering into a symbiotic alliance with mycorrhizal fungi.[28] When conducted in a carbon dioxide-rich atmosphere like that of the Palaeozoic, some 400 million years ago, liverworts colonized by mycorrhizal fungi grew three times larger than their counterparts lacking any mycorrhizal fungal associations, and had a dramatically higher reproductive output. This is what evolutionary biologists like to call the 'fitness' of an organism, and gemmae production jumped four-fold in liverworts with mycorrhizal fungal partners, compared to those plants without them. If such benefits accrued to the pioneers on land, we can imagine that the carbon dioxide-rich atmosphere of the Palaeozoic accelerated the greening of the land by fostering the prodigious production of tiny green emissaries, for establishing satellite populations. Supplying the mycorrhizal fungi with a radioactively labelled source of phosphate demonstrated the transfer of this essential nutrient to the tissues of the host plants. Conversely, supplying liverworts with radioactive carbon dioxide demonstrated that host plants reciprocated by provisioning organic carbon to support fungal growth and proliferation, to produce in the fungus a staggering absorptive area for nutrient uptake from soil, equivalent to that of a tennis court.

The exciting conclusion drawn from these experiments is that partnerships between liverworts and mycorrhizal fungi seem to operate in a 'mutualistic symbiosis' comparable to that found in trees and grasses. At last, Pirozynski and Malloch's hypothesis appeared to have gained crucial experimental verification. Extrapolated to the past, these findings support the notion of proto-bryophyte-type early land plants teaming up with fungi to obtain scarce nourishment from the soil.

Satisfying though all this may be, the story does not end here. In 2011, Martin Bidartondo of the Royal Botanic Gardens, Kew, and his colleagues, made a sensational discovery.[29] Bidartondo is a leading expert in the use of DNA fingerprinting to identify the fungal partners of land plants. Applying his DNA-based probes to strange liverworts collected from New Zealand, Bidartondo and his team, to their surprise, did not find the arbuscular mycorrhizal fungi they expected. Instead, these liverworts turned out to harbour a very ancient and partially saprotrophic fungal lineage (saprotrophs, recall, obtain energy by feeding on decaying matter). This peculiar and primitive fungal lineage is an early-evolving group that includes pea truffles (order Endogonales) and is possibly older than the arbuscular mycorrhizal fungi (order Glomerales).[30] A cautionary note is warranted here because it is difficult to date accurately the early diversification of fungi. Nevertheless, reconstructions based on the best available evidence support the origin of Endogonales fungi as pre-dating that of arbuscular mycorrhizal fungi. What Bidartondo and his team had discovered was that one of the oldest phyla of fungi colonized the oldest lineage of liverworts. Classified within the taxonomic class of Haplomitriopsida, these primitive liverworts represent living relatives of early land plants that descended from freshwater algae by adapting to life out of the water hundreds of millions of years ago. Until the taxonomic confusion about the evolutionary relationships within bryophytes (liverworts, hornworts, and mosses) and vascular plants is resolved,[31] it is difficult to be sure how to interpret the plant–fungus relationship, but it raises the question: could this be the ancestral symbiosis that operated early in plants' tenure on dry land?

Here's the rub, though. If we want to make the argument that this weird partnership between primitive liverworts and fungi facilitated the arrival of terrestrial plant life, we need to know about the functional status of the association. Just because the two groups of organisms live in close association with each other, we cannot assume a harmonious relationship. Tip the balance in favour of the plant and it becomes a commensal relationship, i.e., one organism benefiting without adversely affecting the other. Tip it in favour of the fungus, on the other hand, and the fungus drains resources from the plant and becomes a parasite. Observations of fungi and their intimate connections with the interior of cells of an extant lineage of ancient liverworts are not enough to conclusively establish the status of the association, and the challenge is undertaking experiments with living specimens to find out.

Confined largely to the Southern Hemisphere landmasses, the Haplomitriopsida liverworts that Bidartondo and his colleagues studied have a centre of diversity in Australasia. This geographical distribution suggests a pattern reflecting the far-off days of the Palaeozoic, when the group was widely distributed across the southern supercontinent of Gondwana, which later fragmented into Africa, South America, India, Antarctica, Australia–New Guinea, New Zealand, and New Caledonia. We collected specimens for experiments by visiting New Zealand's South Island, a strange paradise marked out by one of the richest assemblages of evolutionarily ancient land plants to be found anywhere in the world. Consequently, it offers the best modern analogue we have for the earliest stages of the colonization of the land by multicellular plant life. Bryophytes heavily colonize massive exposed rock faces, coating them with a green veneer of photosynthetic life, drip-fed with water draining from the thin infertile soils above. In a scene played out endlessly over millions of years, liverworts, hornworts, and mosses jostle side-by-side for light and what dissolved nutrients there are to be found in the films of water coating the fine-grained rocks.

Our plant-collecting trip to the west coast of South Island was in December, when the keening of the blackbirds in the evenings reminded us it was summertime in the Southern Hemisphere, where the climate, at a latitude of 42°S, is mild and wet. Settlers introduced the blackbirds and other songbirds from Great Britain, as sentimental reminders of home. The rainfall that soaks the western coastline of South Island is generated by the rocky spine of the Southern Alps, which intercepts the westerly winds bringing in moisture-laden clouds from the Tasman Sea. This feeds glaciers, lakes, and wild temperate rainforests, creating ideal habitats for bryophytes to thrive and form close-cropped chlorophyllous carpets in this quiet, unspoilt corner of the world.

English bryologist and retired professor of botany, Jeff Duckett, of the Natural History Museum, London, guided us to collecting localities here and further north, in the pristine, moss-draped, old-growth southern beech (*Nothofagus*) forests. Duckett is usually found dwelling in the Natural History Museum's attic, but for that fortnight he was gasping fresh air as part of the team hunting antipodean bryophytes. In companionable mode, Duckett kept us entertained en route to the collecting sites with tales of botanists getting married underwater, not to mention the curious incident of the three distinguished research professors who holidayed in Iceland, only to be stuck in their hotel after discovering that one was unable to drive, the other would not drive on the right-hand side of the road, and the third only drove on tarmac road surfaces.

The fascinating carpet of bryophytes beneath our feet soon transfixed us, and before long we discovered patches of liverworts belonging to the genus *Haplomitrium* (Haplomitrioposida) (**Figure 20**). These plants have a strange appearance, quite unlike that of their later-evolving liverwort cousins. They grow with sickly yellow subterranean axes that creep through the soil and leaf litter on the forest floor. Occasionally an erect shoot pushes upwards, reaching only a few millimetres or so above the soil surface, wrapped in a series of small, flat, glossy green leaf-like structures.[32] Below ground, the creeping axes are naked, devoid of rhizoids that typically anchor later-evolving liverwort lineages into the soil. Instead, the outer cells of the axis secrete a strange mucous substance. The copious mucilage might act as a water reservoir for the plants to draw on during dry

**Figure 20** *Haplomitrium gibbsiae* (upper picture, between arrows) and *Treubia pygmaeaone* (lower picture), two of the most primitive liverworts on Earth.

spells, or perhaps provision the symbiotic fungus with carbohydrates to sustain it along with the sugars produced by the plant. Inside the tissues of the subterranean stems are distinctive swollen finger-like projections and filamentous coils of pea truffle fungi that extend within cells. Similar fungal projections and coils penetrate the cells of the mucilage-coated underground stems of the other ancient antipodean liverwort lineage, classified to the primordial genus *Treubia*.[33]

Back in the UK, and cultured under environmental conditions mimicking those in New Zealand, our collected specimens of *Haplomitrium* and *Treubia* thrived, making it possible for us to conduct experiments designed to answer key questions about the nature of their alliance with pea truffle fungi.[34] We fed the plants radioactive carbon dioxide that was converted subsequently into radioactively labelled organic compounds via photosynthesis. This meant we could determine whether the plants delivered these sugars to the fungi. Fungi were, in turn, supplied with specific sources of phosphorus and organic nitrogen, each labelled with radioactive forms to allow traceability. From this work, we discovered the fungi were adept at capturing and transferring nutrients from soil to their host liverworts. When we conducted the experiments under the simulated carbon dioxide-rich atmosphere of the Palaeozoic, the fungi gained greater supplies of carbon from their hosts. Could this be a clue to explaining why they switched from a saprotrophic lifestyle, involving feeding on dead tissue, to a mutualistic relationship with living plants? By forming associations with plants, they could obtain abundant and reliable supplies of photosynthate in return for assisting the uptake by the plant of growth-limiting nutrients. Complicated experiments like these ultimately provided evidence that Haplomitriopsida liverworts and pea truffle fungi are engaged in a mutually beneficial symbiosis.

Still, mycologists are a difficult bunch to satisfy and demand a high burden of proof if you want to demonstrate that the two groups of organisms function in a symbiosis. For them, a prerequisite to establishing the functional basis of any relationship between a fungus (and indeed any microorganism) and its host plant is that the putative partners be grown separately, then brought together to determine whether the naturally occurring symptoms and structures of the association are reproduced. This procedure is one of Koch's Postulates, after the German physician Robert Koch (1843–1910), who first proposed what has become central to the diagnosis of pathogenesis and the aetiology of disease. Working with specimens collected from New Zealand, we successfully cultured *Haplomitrium* and *Treubia* and the pea truffle fungi separately in sterile petri dishes laced with sugary agar-gel. Without their fungal associates, liverworts failed to develop normal leafless axes and failed to produce copious mucilage, suggesting fungal partners induce this feature of their biology. Without their plant partners providing them with a supply of photosynthate energy, the fungi grow miserably (if at all) relying on their saprotrophic capabilities to obtain energy. Re-introducing them to each other induces substantial anatomical changes in the host plants. Liverworts in combination with their fungal partners reproduced the typical symptoms of colonization seen in the specimens collected from field sites on South Island.[35]

How might these findings from laboratory experiments fit in with remote events back in the Ordovician? A curious feature of the biology of pea truffle fungi is that some of them can live a dual lifestyle, functioning either as free-living saprotrophs or as symbionts teamed up with plants. Given this versatility, it is possible these fungi lived on the Ordovician landscape long before terrestrial plants arrived on the scene. Functioning in saprotrophic mode, they could have made a living by obtaining nutrients from decomposing organic matter formed from algal and cyanobacteria remains on the margins of shallow freshwater ponds and lagoons. When land plants appeared, the fungi could have seized the ecological opportunity to establish a new symbiotic alliance with a wonderful, new, reliable energy source for doing what organisms like to do: grow and reproduce. Plant life benefited from the arrangement by securing access to soil nutrients from the fungi, including inorganic nutrients like phosphorus and organic nitrogen from decaying litter and detritus. We can see now that bringing the pea truffle fungi into the evolutionary picture seems to offer a solution to the unlikely coincidence of two distinct groups of organisms (plants and fungi) making land more or less simultaneously.

Notice that only an experimental approach to fundamental questions raised by the fossil record can establish essential functional evidence supporting or refuting the long-standing conjecture of the presumed symbiotic origins of our land floras. Adopting such experimental approaches to unpicking symbioses has its roots in the philosophy of Anton de Bary (1831–1888). De Bary is widely credited with coining the term symbiosis, and strongly urged his fellow mycologists to undertake experiments to determine the functional nature of symbioses between plants and fungi. Some might argue that experiments like the ones we have been discussing for early land-plant lineages necessarily employ contemporary fungal strains, and might question the wisdom of drawing inferences from such studies. Genomics provides the counter-argument. Fungal genes controlling functions of the symbiosis are ancient and highly conserved, retaining similarities to fungi in their far more distant ancestors from which they diverged hundreds of millions of years ago. Similarly, the genetic toolbox plants required for the establishment and functioning of the symbiosis dates back hundreds of millions of years to a common mycorrhizal ancestor.[36] On this basis at least, experimental findings offer insights into the past, the product of a marriage between contemporary ecology and palaeontology.

Not surprisingly, a resurgence of interest in fossil fungi followed, with palaeontologists re-examining thousands of thin section slides cut from the Rhynie Chert fossils in search of evidence for early symbiotic encounters.[37] One reward of this effort has been the discovery of fungal symbioses in the early Devonian fossil plant *Horneophyton lignieri*, which has cells with features suggestive of both arbuscular mycorrhizal and pea truffle fungi.[38] The evidence is far from clear-cut, but the intriguing suggestion is that the plant benefited from a dual fungal symbiosis. This suggestion fits with recent studies showing examples of liverworts, hornworts, lycophytes, and ferns entering into dual fungal partnerships, although the functional purpose of such a strategy remains unclear. Did early land-plant lineages, like their living descendants, associate with a wider and more versatile suite of fungal symbionts than previously assumed, interacting in ways still to be discovered?

A case in point arose from the genetic fingerprinting of the fungal partners of surviving early vascular land-plant lineages, which offered a glimpse into what we may have been missing. The plants concerned were ancient grades of spore-bearing vascular land plants including the lycophytes and primitive ferns like adder's-tongue (*Ophioglossum*) and moonworts (*Botrychium*) that are descended from some of the earliest known vascular land plants.[39] The life histories of these plants fascinated W.H. Lang, who reported discoveries of fossilized Devonian terrestrial life at Rhynie. Lang was quick to appreciate what our modern flora can contribute to understanding past evolutionary events.[40] In the simple vascular land-plant groups, spores shed by adults germinate into diminutive plants (gametophytes) that are essentially small packets of cells, a few millimetres in diameter. Buried in soil and thick layers of leaf litter, the developing plants are a distinctive phase of the life cycle that lack chlorophyll and are unable to photosynthesize. Without being able to feed themselves by photosynthesis, it is easy to be fooled into thinking these small subterranean plants are making a living as saprotrophs, obtaining nourishment directly from decomposing leaf mould in the soil.[41]

The truth is that the romantic notion of saprotrophic plants is a myth, decisively debunked by Leake, whose forensic dissection of the literature shredded the supposed supporting evidence.[42] Clues to the unusual mode of nutrition supporting their growth surfaced when detailed microscopic observations revealed them to be colonized by fungi, which later DNA analyses identified as mycorrhizal fungi.[43] Instead of feeding themselves by photosynthesis, or acting as saprotrophs, these tiny plants hidden underground are supported by carbon energy subsidies supplied by mycorrhizal fungi. It is an unusual mode of nutrition and a remarkable case of plants parasitizing fungi. Now recognized in over 400 species, this way of life (known as myco-heterotrophy) evolved many times in different groups of plants, possibly as an evolutionary solution to living in the shaded forest understory where effective photosynthesis is difficult.[44]

The story of plants parasitizing fungi took an intriguing new twist when researchers at the University of Colorado discovered that subterranean gametophytes and photosynthetic sporophytes of several species of lycophytes and ferns associate with the same specific group of mycorrhizal fungi.[45] The startling implication here is that adults (sporophytes) photosynthesizing in the sunlight above ground are feeding offspring (gametophytes) buried in soil with carbon supplied through shared networks of mycorrhizal fungi.[46] In all, about 100 species belonging to several genera of early vascular land plants (*Lycopodium, Huperzia, Psilotum, Botrychium,* and *Ophioglossum*) produce gametophytes that develop in soil or litter and lack chlorophyll. Unable to feed themselves by photosynthesis, they are entirely dependent on subterranean mycorrhizal networks connected to photosynthetic adults to supply carbon and nutrients. The idea is still in its infancy but it is a remarkable thought that green photosynthetic adults are nurturing progeny buried in soil via intergenerational carbon subsidies delivered through common networks of mycorrhizal fungi.[47] If correct, the situation amounts to a case of parental nurture by adults of the same species above ground. Given the antiquity of the plant groups involved, parental nurture through shared fungal symbiotic partners may have been a previously unrecognized feature of their ecology for millions of years.

Consider how it might work over the life cycle of the adder's-tongue fern (*Ophioglossum*). As the spore germinates, it develops into a gametophyte fuelled by carbon energy supplied by photosynthetic adults above ground and delivered by its fungal partners. Following fertilization, the young sporophyte slowly develops by producing one or two small green photosynthetic leaves that emerge from the soil. Such an arrangement has the advantage of enabling sexual reproduction to occur underground, with the gametophyte buried in the moist soil and afforded protection from desiccation.[48] Thanks to relying on carbon subsidies from a mycorrhizal network to support their young, adults can produce prodigious numbers of dust-like spores. If you have ever discovered ferns such as adder's tongue in pastures and dune slacks, then you will appreciate that they often cluster together in large patches, containing hundreds of individuals. Conceivably, in such circumstances, adults are nurturing the youngsters of neighbouring plants via common mycorrhizal networks, and this arrangement amounts to a form of co-operative breeding in land plants.

It is an extraordinary possibility and wrests the concepts of parental investment and co-operative breeding away from the clutches of my zoological colleagues along the corridor and into the hands of botanists. Ben Hatchwell, professor of behavioural ecology at the University of Sheffield, has made his name working out the details of co-operative breeding in long-tailed tits (*Aegithalos caudatus*). These fascinating and highly social little birds are often heard before they are seen, flitting through gardens in flocks in the wintertime. During the spring, males and females pair up and work together to build intricate bottle-shaped nests with a hole at the top. If disaster strikes, and the chicks are eaten by a predator, the parents may abandon the nest and switch their attention to helping feed the chicks of a relative. This turn of events (which may seem surprisingly altruistic at first glance) sees the survival rates of these chicks improve thanks to 'helper' birds raising the family. Extended family life continues after the fledglings become independent, with both parents and helpers investing time and effort for weeks afterwards to ensure survival. In long-tailed tits, Hatchwell has observed that individuals are more likely to help feed the chicks in nests belonging to a close relative rather than distant relatives or unrelated individuals.[49] Helpers are effectively improving the survival chances of their genes indirectly, even though their own attempts to pass on their genes directly by breeding have failed. We can marshal similar arguments for the lycophytes and 'lower' ferns, which grow in patches of closely related adults raising the offspring of close relatives to ensure the survival of their genes.

If the notion of parental nurture of progeny via common mycorrhizal networks in the plant world seems far-fetched, we should note that there may be a precedent for it in the orchids. Orchids are the most advanced group of plants on Earth and, with over 27 000 species, also the most diverse. They reproduce by releasing prodigious quantities of, typically, wind-dispersed dust-like seeds. Acting as tiny emissaries for the colonization of new habitats, the seeds require a symbiotic fungal partner for the provision of resources to germinate successfully and develop into a seedling. This strategy allows orchids, like the more ancient groups of lycophytes and ferns, to maximize seed production and wind dispersal through minimal maternal investment. Parent plants rely on symbiotic fungi to make good the deficiency in carbohydrates needed for seedling establishment.[50] The analogy between advanced and primitive vascular land plants extends further. Subterranean orchid seedlings lack chlorophyll and form tuber-like structures that so closely resemble the non-photosynthetic gametophytes of lycophytes and ferns they share the same unique botanical term—a protocorm (**Figure 21**). Striking anatomical similarities exist among the structures of the protocorms of these widely separated evolutionary groups of plants (orchids and lycophytes), which associate with very different groups of symbiotic fungi, to secure and sequester carbon supplies obtained from them.

All this was unknown to Charles Darwin, who was fascinated by orchids and their spectacular floral diversity and published his classic work on this group, *The various contrivances by which orchids are fertilized by insects* in 1862. Not a trifling effort but a great labour of love, the book was written after 20 years of studying every facet of orchid biology. Yet germinating orchid seed proved beyond him. In frustration, Darwin wrote a letter in 1863 to his friend and mentor Joseph Hooker, Director of the Royal Botanic Gardens Kew, saying that despite having 'not a fact to go on', he had 'a notion, (no, I have a firm conviction) that they [germinating orchid seeds] are parasites in early youth on cryptograms [i.e., fungi]'.[51] Darwin was right, and ahead of his time. For the crucial involvement of fungi in orchid

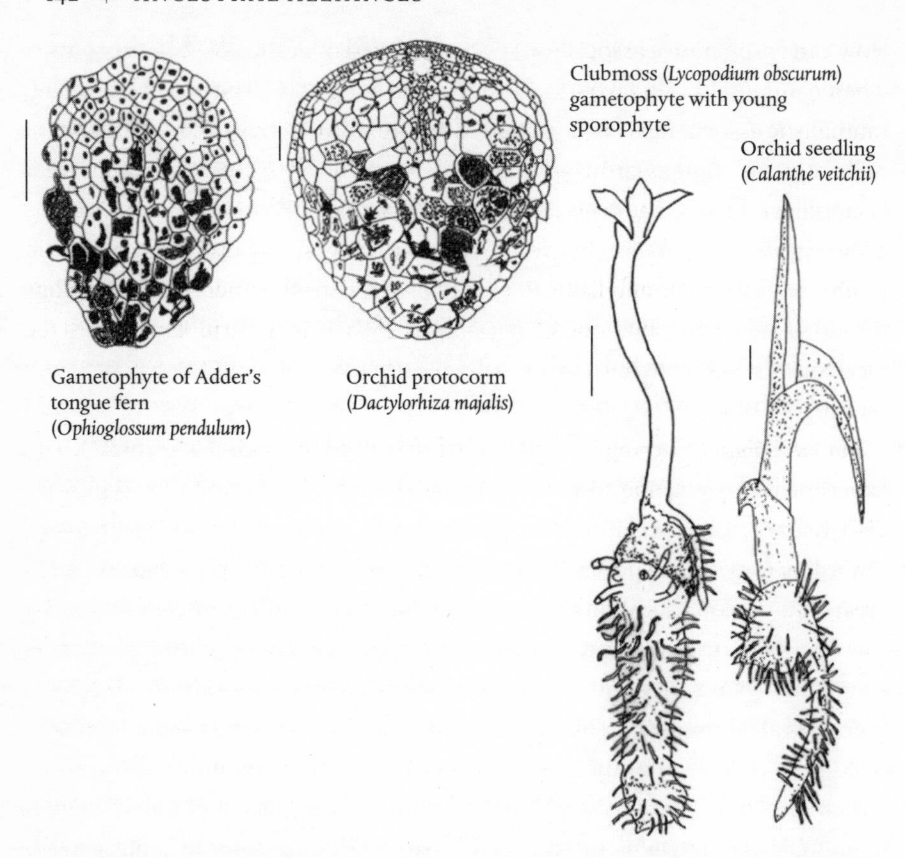


**Figure 21** The soil-dwelling gametophyte of a fern bears a striking structural resemblance to the protocorm of a highly evolved orchid, as does the young sporophyte of a clubmoss and an orchid seedling. Scale bars = 100 mm in all diagrams except *Lycopodium obscurum* where it = 500 mm.

seed germination was not discovered for another three decades, when it was reported in 1899 by the French botanist Noël Bernard (1874–1911).[52] Darwin gave no reason for his 'firm conviction'. Presumably, the tiny size of orchid seeds raised the possibility in his mind that they required additional food reserves from some other source (or organisms) to ensure successful germination—a situation that obviously parallels the dust-like spores of lycophytes and ferns, early-evolving vascular plant lineages.

Darwin recognized that mutualistic associations like those between fungi and land plants apparently presented a major problem for his evolutionary theory.

How can co-operative associations between different groups of organism persist when natural selection favours selfishness and conflict? Despite his puzzlement, mutualistic associations are common in nature from fig trees and their fig-wasp pollinators, to fungus-growing ants, and to squids with their bioluminescent bacterial symbioses. The central evolutionary paradox of mutualistic associations is the uneasy co-operation between organisms. Uneasy because the relationship is inherently unstable and liable to break down if either partner selfishly exploits the other by taking but not giving anything back in return. There must be mechanisms for stabilizing it in the long run; otherwise, how could it have endured for millions of years?

For land plants entering into mutualistic partnerships with mycorrhizal fungi, experiments are starting to reveal some answers to these questions. Plants, we discover, are very discriminating in their choice of mycorrhizal fungal partner, rewarding only those species that provide a steady supply of phosphorus with a greater supply of photosynthate, and sanctioning less co-operative partners by shutting down their carbon supply.[53] The reciprocal arrangement is that roots supply more carbon to the fungus, which obtains more phosphorus in return. A continual reinforcement of this sort helps stabilize the partnership and may be one of the reasons why the symbiosis between plants and mycorrhizal fungi has endured for hundreds of millions of years. It is distinct from other examples of mutualism found in nature, where one partner often appears in control. In the mycorrhizal mutualism, both sides—plants and fungi—interact with multiple partners, making each resistant to being enslaved. Nature's free market economy suggests an uneasy co-operation stabilized by competitive partners on both sides, remunerating those supplying the highest quality services.[54]

Stability does not necessarily mean there are no freeloaders. 'Cheater' plant species lacking the ability to photosynthesize can break co-operative arrangements to become parasites. Sir David Read at Sheffield University uncovered a fascinating example of this phenomenon with the case of a liverwort 'cheating' on a symbiotic partnership between trees and their fungal partners by 'stealing' carbon. The cheat in question is a strange thalloid liverwort called ghostwort (*Aneura mirabilis*) that lives in the litter and soils beneath European birch, pine, and oak forests and has lost the ability to photosynthesize. Ghostwort is the only known example of a non-photosynthetic bryophyte. It lives a subterranean lifestyle and is dependent on symbiotic fungi for supplies of energizing carbon compounds and other nutrients, which it obtains by cheating—it parasitizes the symbiosis by tapping into a fungal 'internet' linking the roots of forest trees to their symbiotic fungi. In essence, these liverworts are stealing the carbon they need from the trees in the network rather than obtaining it from the soil itself.[55] In an evolutionary sense, *Aneura* appears to have 'learned' which clade of mycorrhizal fungus is a soft touch and can be tricked out of its persistent supplies of carbon delivered by the host tree. *Aneura* exhibits extreme fungal fidelity, only associating with a specific fungal clade that forms partnerships with pine and birch trees. How it manages to be so highly selective in forming this specific association rather than exploit those fungi connected to the roots of other tree species is an open question. Molecular geneticists soon became interested in the unusual story of this tripartite symbiosis between a tree, a fungus, and a liverwort.[56] They discovered that exploiting fungal carbon sources relieves the liverwort of the need for the genes required for photosynthesis. And this seems to be rule rather than the exception for parasitic plants. The chloroplasts of the flowering plant species *Epifagus virginiana*, which is strictly parasitic on the roots of beech trees, have also lost the ability to photosynthesize, and with it, a similar number of genes.[57]

Over 40 years ago, Pirozynski and Malloch presciently argued that the Devonian fossil record of plant life raised fundamental questions concerning the symbiotic origin of plant life on land. They provocatively raised the intriguing hypothesis that an ancient partnership between symbiotic fungi and land plants was essential to colonizing the continents. It challenged future generations to evaluate critically their speculative idea. Recent discoveries are giving greater clarity and offering new perspectives on the sequence of events that lay behind the origin of the terrestrial biosphere and the identity and roles of the fungal groups involved. Matters that had remained speculative for nearly half a century are being resolved, crystallizing a more complex and fascinating picture than Pirozynski and Malloch could have envisaged.

Our best reconstruction for the earliest events is that primitive fungi formed part of early soil-forming communities in the Ordovician landscape, over 420 million years ago. Some of these lineages included the pea truffles and their close relatives, which operated in partially saprotrophic mode, feeding on decaying algae and cyanobacterial mats. As simple rootless early plant life made the transition to land, beneficial plant–microbe alliances developed. With plants offering a safe, carbon-energy-rich haven, some lineages of pea truffle fungi abandoned their saprotrophic lifestyle to forge symbiotic alliances with them. Plants, in return, gained access to soil nutrients captured by their fungal partners. As plant life on land evolved, and diversified, the first humus-rich soils developed and the economics of earlier arrangements shifted. Plants developed symbiotic partnerships with arbuscular mycorrhizal fungi specially adapted for mining and extracting elemental nutrients from rocks and minerals, especially phosphorus—an essential ingredient for building crucial biomolecules needed to run their metabolism. Some pea truffle fungal groups reneged on the contract, abandoned the plants that had served them so well, and went back to decomposing the increasingly abundant organic matter to obtain energy. Other fungal groups stayed put, allowing some plants to benefit by collaborating with two different groups.

The co-evolution of plants and symbiotic soil fungi continued, and in the next phase a new type of mycorrhizal association appeared in which ectomycorrhizal fungi began colonizing tree roots. The oldest fossils are of roots belonging to Pinaceae trees, and the tropical tree group Dipterocarpaceae, and date to around 41–55 million years ago, although evidence from molecular clocks hints at much earlier origins during the Jurassic, about 180 million years ago. Ectomycorrhizal fungi evolved multiple times from saprotrophic ancestors and most reproduce sexually by developing fruiting bodies above ground—recognizable mushrooms like bolets, amanitas, and chanterelles. They associate mainly with the roots of boreal and temperate tree species, and support the health of forests by releasing enzymes to digest humus in the soil and extract nutrients otherwise not available, then transfer them to the trees.

The sequencing of fungal genomes has shown how ectomycorrhizal fungi have shed unwanted genomic baggage inherited from their saprotrophic ancestors to make their symbiotic lifestyles possible.[58] 'Unwanted' because saprotrophic fungi decompose dead plant tissue in soil litter for energy by secreting enzymes that chemically break down the tough plant cell walls, and fungal pathogens use similar enzymes to digest or rot living tissues. This makes it difficult, if not impossible, for both fungal groups (saprotrophs and pathogens) to enter into a symbiotic relationship with plants, because the enzymes would damage the host and elicit plant defence responses. By shedding whole families of genes encoding for the enzymes that saprotrophic and pathogenic fungi use to degrade cell walls aggressively, ectomycorrhizal fungi enjoy a beneficial symbiosis with trees. Nevertheless, they still possess a small collection of genes encoding selected enzymes for mobilizing carbon and nitrogen from well-decomposed organic matter to help prospecting fungal filaments survive in soil litter when not in engaged in a symbiosis with roots.

Most recently (geologically speaking), around fifty million years ago, in an extraordinary twist to the symbiosis story, legumes such as peas and beans co-opted the symbiotic genetic machinery for establishing partnerships between plants and fungi to develop a new sort of symbiosis with bacteria. These plants teamed up with specialized bacteria able to perform the remarkable metabolic trick of converting atmospheric nitrogen into a form plants can use. Specialized structures called root nodules host the symbiosis. These are the familiar white, often pinkish, swellings found around the roots of beans, peas, and other members of the legume family (Fabaceae or Leguminosae). The genetic toolkit for establishing the nitrogen-fixing bacterial symbiosis inside the nodules derives from the pre-existing machinery used for establishing the symbiosis with arbuscular mycorrhizal fungi. Sequencing of the genome of the legume *Medicago truncatula* revealed a whole genome duplication event, which occurred approximately 58 million years ago,[59] that duplicated the mycorrhizal symbiosis toolkit genes. This seems to have created the gene copies necessary for nodulation and the nitrogen-fixation symbiosis to evolve. Once again, genome duplication provided spare genetic material to enable evolutionary development, in this case a major new plant–microbe partnership—the bacterial symbiosis.

Putting all this together, and adding a little more detail to fill in the gaps, allows us to sketch out a scenario for the evolution of plant–microbe symbiotic alliances as the continents turned green. Curiously, though, the fossil record offers rather little on the possible involvement of root–fungal associations as primitive forest ecosystems evolved and expanded on land during the Devonian. But by Carboniferous times the mycorrhizal symbiosis had diversified, with the fungi colonizing the fine rootlets of the rhizomes of the giant arborescent lycopsids that dominated the extensive low-latitude swamplands. They also colonized the roots of lycophytes, and the early relatives of conifers, that formed large forests elsewhere on the continents at this time, alongside the seed ferns. On they marched, as their opportunistic spread into the plant kingdom continued by later colonizing the roots of some gymnosperms (e.g., cycads, conifers) that dominated the land flora between the Triassic and Cretaceous. Sometime later, in the Jurassic and Cretaceous, the first ectomycorrhizal associations appeared. Finally, around the Eocene Epoch, along came the nitrogen-fixing symbiosis between plants with bacteria (**Figure 22**).

This emerging story of the evolving symbiotic lives of plants and microbes, based mostly on fossil evidence, is also supported by evidence from molecular biology. Discoveries in the field of molecular genetics have revealed that widely divergent host plant species use strikingly similar genetic toolkits to establish symbioses with fungi or bacteria. The implication is that plants share an evolutionary origin for this ancient symbiotic pathway that was inherited from a single common mycorrhizal ancestor that lived hundreds of millions of years ago.[60] In fact, not only do land plants share common genetic symbiotic machinery but some of those components were inherited from an algal ancestor.[61] This pushes the origin of the symbiosis toolkit back to the most recent common ancestor of extant land plants and green algae. That lineage of algae, for some reason, seemed to be pre-adapted for interacting with beneficial fungi, and later facilitated the colonization of the continents by plant life.

If this all sounds like tales woven from the ivory towers of academia with little relevance to modern society, we need a reality check. Finding alternatives to the production of nitrogen fertilizers is essential for sustainable food production. The roots of cereals establish arbuscular mycorrhizal partnerships and discoveries of how the nitrogen-fixing symbiosis operates in legumes raise the possibility of engineering it into crops. Can we co-opt the pathway between mycorrhizal fungi and the roots of cereal crops to allow the roots to host nitrogen-fixing bacteria? This would solve the problem of pollution caused by overuse of artificial nitrogen fertilizers, and address shortages and high costs of fertilizers in developing countries, particularly sub-Saharan Africa.[62] The first step to enabling crops to make their own fertilizer is to get cereals to exchange molecular crosstalk with nitrogen-fixing bacteria by releasing substances like flavonoids. Step two is far harder: convince cereals to host the bacteria in their roots. Clues to the way forward may come from understanding how mycorrhizal fungi evolved root colonization strategies to trick their way past plant defences by circumventing plant immune responses.[63]

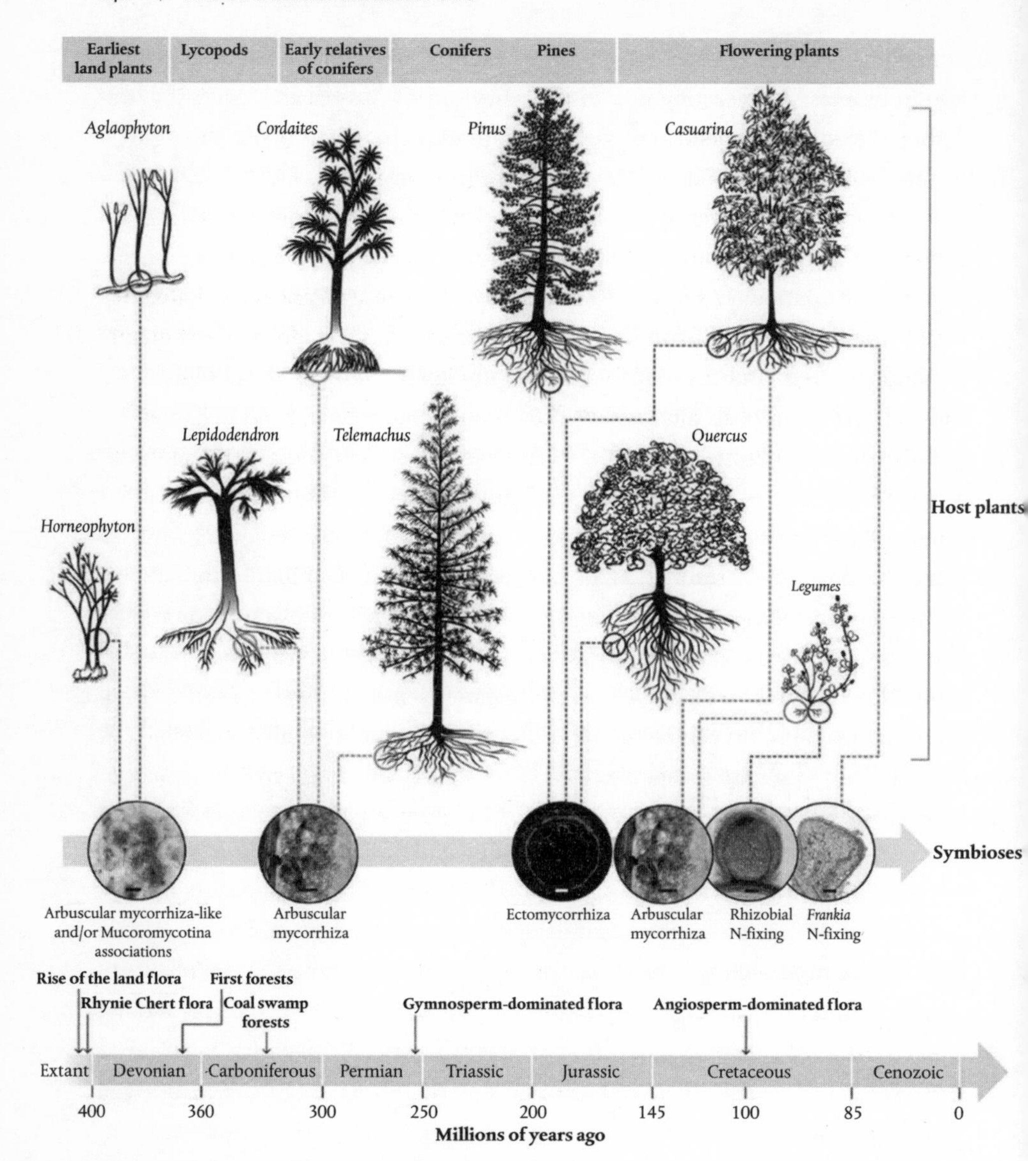


**Figure 22** A speculative scenario for the evolution of beneficial alliances between plants and microbes over the past 400 million years.

Meanwhile, forests create a huge carbon 'sink' as they remove carbon dioxide from the air to build leaves, wood, and roots, and to feed mycorrhizal fungi. Although the sink varies from year to year, on average it absorbs one-quarter of our carbon dioxide emissions from burning fossil fuels, with the forests helped by mycorrhizal associations. They allow forests to overcome soil nutrient limitations to promote carbon dioxide fertilization of plant growth and store carbon in biomass. Forecasting the future behaviour of this natural forest sink for carbon dioxide will depend in part on incorporating the effects of these mutualistic microbes into our models of the global carbon cycle.[64]

We have seen how, over the great sweep of Earth's history, from the Ordovician ~470 million years ago onwards, soil-borne microbial symbionts played a crucial role in the evolutionary success of the plant world. These alliances facilitated the establishment of plant life on land, with major ramifications for understanding the past, present, and future ecology of complex land-based ecosystems. Had Kidston and Lang been alive today, they would surely have been delighted to know how their lifetime's work is inspiring new directions and new generations of scientists. Lang, in particular, felt a 'plant is a very mysterious and wonderful thing and our business as botanists is to try and understand it as a whole'.[65] This admirable perspective resonates with our central theme of symbiotic associations as evolutionary enablers promoting the amazing diversity of plant life on Earth. Margulis puts the point more poetically. In her mind, this symbiosis 'was the moon that pulled the tide of life from its oceanic depths to dry land and up into the air'. In Chapter Seven we discover how, as plant–fungal alliances evolved, turning the continents green, plants also began slowly, imperceptibly, to sculpt the climate. The era of land plants as 'bioengineers' of planetary climate had begun.

# 7 SCULPTING CLIMATE

'The most recent dream of the New York State Museum was realized when on February 12, 1925, there was formally opened to the public the restoration of the extensive forests that flourished in eastern New York a few hundred million years ago.'

Winifred Goldring, *The Oldest Known Petrified Forest*, 1927

In the 1850s, Minister Samuel Lockwood (1819–1894) discovered a remarkable sandstone cast of a fossil tree stump near the former settlement town of Gilboa (from the Hebrew for boiling springs) in New York State.[1] Lockwood had already found dinosaur fossils in the area but his fossil stump was special. It proved to be the first documented discovery of an ancient fossil tree in North America—a palaeobotanical taster of things to come, because in 1869 workers quarrying rock for repairing flood-damaged roads and bridges discovered fossil tree stumps rooted in the position in which the trees originally grew. A few decades later, another set of tree stump casts turned up in Riverside Quarry, Gilboa. Resembling Apollo spacecraft capsules, each sedimentary cast measured the span of your hand at the neck, flaring to two or three times that at the base.

The casts are really voids, created by the decaying tree stump, that became back-filled with sediment. Gathered together by palaeontologists, they formed the start of the Gilboa fossil plant collections. Winifred Goldring (1888–1971) (**Figure 23**), the first woman to hold the position of State Palaeontologist at the New York State Museum, later constructed the innovative Gilboa forest exhibit documenting 'the world's oldest forests'.[2] Neatly arranged outside the little weatherboarded museum on the outskirts of Gilboa sit the same casts on their bed of gravel, and inside you can find the story of their discovery (**Plate** 7).

**Figure 23** Winifred Goldring (1888–1971) in the field at Rensselaer Falls, New York, in 1939, the year she was named State Paleontologist of New York, a position she held until 1954.

Goldring published formal scientific descriptions and illustrations in 1927, and concerted efforts nearly a century later by the palaeobotanists Chris Berry, Bill Stein, and their colleagues have transformed the scientific picture of these ancient forests that once flourished near the shoreline of an inland sea. They unearthed spectacular slender fossilized trunks and crowns of specimens belonging to extinct trees with the catchy name of cladoxylopsids, known previously only from those sedimentary stumps.[3] The reconstruction of trees from such finds is a difficult task, but Stein and Berry are masters in the art of piecing together the fragmentary remains. They managed to assemble the entire structure of these unusual 8-metre tall trees by reuniting the crowns and trunks with the bases. The cladoxylopsid trees left behind root mounds forming circular craters with a diameter about the size of a football.[4] Each depression is really a counterpart mould of the tree preserved as a sediment cast. Anchored into the soil by shallow rootlets that sprouted from the bulbous base of their trunks, the trees had a similar arrangement of strap-like rootlets to that found around the base of tree ferns that are often planted in the glasshouses of botanic gardens. Tree ferns share a similar basic body plan with those early trees, producing a whorl of feathery branches unfolding from a central growing region located on top of a stout trunk. The first Devonian giants of the plant kingdom achieved their great stature with a similar economy of form, but, unlike tree ferns, these strange trees grew without leaves. Instead, they photosynthesized with a crown of short-lived branches arranged geometrically above a single long-lived trunk reaching a height of several metres.

The fossil trees of the early forests found in rocks near Gilboa date back ~385 million years to the Devonian. This was the remarkable time of explosive evolutionary action when plants evolved from small primitive forms that barely reached your ankle into large, complex trees. A spectacular variety of tree-sized photosynthesizers soon appeared and became progressively larger and complex within ~20 million years. Scrambling forms sprouting from rhizomes appeared alongside the cladoxylopsids, followed by progymnosperms, the forerunners of conifers, made of woody trunks topped with leafy canopies and secured into soils with woody rooting systems. These evolutionary events in the arboreal realm changed the world forever, altering the course of Earth's climate history by entraining a web of planetary feedback processes that reshaped the chemistry of the oceans and atmosphere.

Over twenty years ago, the pioneering Yale geochemist Robert Berner (1935–2015) began unpicking linkages in the complex chain of cause and effect that connects trees and Devonian global climate change. His interest in trees began 'in the early 90s while sitting on the banks of the Potomac River, upriver from Washington, D.C., [when he] noticed plants apparently growing out of rocks. I had seen this before and at that time I became interested in how plants affect rock weathering', before remarking 'besides being an interesting phenomenon in its own right, the study of plants and how they affect the rate of weathering is important in the long-term carbon cycle'.[5] By 'how plants affect rock weathering', he means how plants help chemically and physically break down rocks that then react with carbon dioxide. The process has important consequences for climate because over millions of years it slowly removes carbon dioxide from the atmosphere, weakening the greenhouse effect and cooling the planet. We shall unpack the details shortly. By long-term carbon cycle, Berner is referring to the slow transfer of carbon between rocks, the oceans, and the atmosphere. Acknowledging his debt to earlier scholars from the 1950s, his subsequent series of scientific papers took geochemists, botanists, and anyone else prepared to listen by the hand and opened their eyes, explaining how Earth's transition from a naked to a forested planet revolutionized its climate history.[6]

Struck by his remarkable view of the place of trees in our planet's climate history, in the spring of 1997 we invited Berner to speak at a scientific meeting organized at the Royal Society, London.[7] Breaking all the usual rules of engagement we teach to graduate students, Berner stood with his back to the audience, facing the screen displaying his presentation, and gave a straight-talking narrative describing his scientific journey towards this new world view. Central to the talk were his discoveries about how trees accelerate the chemical weathering of basalt rocks in the Skorradalur Valley, south-western Iceland.[8] By analysing the chemistry of stream water draining from small catchments populated with stunted birch/pine trees, or barren rocks colonized by patches of lichens and mosses,[9] he revealed the distinctive geochemical fingerprint of trees chemically dissolving basaltic rocks. Berner's thinking was that the tree-populated catchment represented an approximate analogue for early forested Earth, and the barren catchment a kind of 'control' for naked Earth; it had lost its trees sometime in the last century from overgrazing by sheep.[10] If these findings can be taken as any sort of guide to chemical weathering regimes before and after the evolution of trees (pre- and post-Devonian worlds, in other words), they suggest that the arrival of trees on the evolutionary stage accelerated rock weathering by a factor of five. Subsequent interest in Berner's results prompted similar studies by other scientific teams and, in the years that followed, they found that no matter which part of the world you are in, the activities of plants accelerate the chemical destruction of Earth's rocky landscape.[11] Berner had discovered a profound truth about how the world works: the activities of forests and trees regulate a fundamentally important process—rock weathering—that controls carbon dioxide levels in the atmosphere and climate over millions of years. Trees were implicated as bioengineers of Devonian global change.

The connections leading to this conceptual advance begin with the quest by roots, and their associated microbial partners, for nutrients, such as phosphorus and potassium, to support their growth. To mobilize these elements they deploy a repertoire of metabolic tricks that acidify soils, chemically liberating the cargo of calcium and other elements from rocks and minerals. Indeed, the distinctive chemical signature of trees that Berner had detected in stream water in Iceland was a higher concentration of calcium and other elements compared to that in the stream water draining through his 'naked Earth' control catchment. These weathered products are transported to the sea by rivers where they react with carbon dioxide dissolved in seawater to slowly precipitate as solid calcium carbonate on the seafloor or become incorporated into the carbonate shells of organisms, such as corals, shellfish, and microscopic plankton called foraminifera. When they die, their shells sink in a continuous rain of carbonate debris that accumulates on the seabed. Slowly over millions of years, thick layers of carbonates build up on the floors of the world's oceans, quietly sequestering carbon that was once in the atmosphere. The long-term sequestration of atmospheric carbon dioxide in this way weakens the greenhouse effect and slowly cools the planet. Eventually, hundreds of millions of years later, this carbon is recycled when plate tectonics folds the seafloor down into the crust. There the carbonates are subject to immense heating and compression on geological timescales of millions of years, converting them back into carbon dioxide gas, released by volcanoes into the atmosphere to continue its endless cycling around the planet.[12]

Meanwhile, mindful of Berner's successful scientific insights gained from working in Iceland, others later visited its basaltic terrain in the hope it would give up the secrets of how early land plants interacted with the planet's rocky landscape millions of years before the trees took centre stage in the Devonian drama. The idea was to travel as far back in time as the Ordovician, over 450 million years ago, to the days when multicellular photosynthetic plants first arrived on land, and extract shallow rock cores, a few centimetres deep, beneath communities of bryophytes, fungi, and algae. Scrutinizing these mini rock cores with a suite of state-of-the-art geochemistry techniques produced little evidence for serious chemical alteration below depths of a few millimetres.[13] At best, a few rock grains beneath liverworts showed features indicative of weathering, including tiny bowl-shaped pits and thin chemically altered weathering rinds, which were cautiously attributed to activities of the plants and their microbial partners.[14] Based on this evidence, communities of simple land plants living in the early Palaeozoic failed to change the world.[15] That development awaited the evolution of trees and the emergence of forests.

Or did it? Some scientists have speculated boldly that the first simple land plants were a 'geological force of nature' right from the off in the Ordovician.[16] At that time, big landmasses of the world, such as Australia, Africa, and South America, were sutured together to form the supercontinent of Gondwana, situated near the South Pole, and the atmosphere contained twenty times as much carbon dioxide as today. Yet there was a major glaciation in the Late Ordovician. Even after accounting for the reduced output from a younger Sun that burned less brightly in Ordovician times, it is difficult to trigger glaciations without drawing substantial amounts of carbon dioxide out of the atmosphere. Did scattered patches of early land plants clinging to rocks dramatically enhance weathering to pull enormous amounts of carbon dioxide out of the atmosphere and trigger glaciation? Did accelerated weathering on land repeatedly flush nutrients in shallow coastal waters to fuel algal blooms? If so, this might explain transient episodes in which burial of organic matter in the oceans increased, thought to reflect ancient pulses of marine productivity. Promising though it may seem at first glance, the flaw is that the shallow rhizoid-based anchorage systems of early land floras limited their zone of influence in the rocky landscape, as the studies in Iceland, and elsewhere, have shown.[17] In some circumstances, lichens can actually *protect* rocks from the climate by smothering the surfaces and slowing down weathering.[18] Not surprisingly, the jury is still out on whether diminutive rootless floras drove the massive changes in global biogeochemical cycling necessary to trigger ice ages in Ordovician times.

Contrast this with the situation that we think unfolded as large photosynthesizing trees and forests evolved and spread across the continents, marching into upland regions during the Devonian as the drive to capture sunlight took hold. Truly large plants with the arborescence habit had arrived. Massive tree trunks topped with stout branches adorned with leaves and leaf-like filaments emerged from a crowded forest floor (**Plate 8**). Trees went from reaching heights of a metre or so to over 30 metres within a few millions of years. The geological periods of the Devonian, and then the Carboniferous, saw the world populated with large trees as they stole a march on terrestrial floras from the tropics to the sub-arctic. Invoking the linkages discussed earlier, suggests that the rise and spread of trees forming the burgeoning forests locked Earth into a set of feedbacks along the pathway of cause and effect: bigger trees → deeper roots → more weathering → less carbon dioxide in the atmosphere → cooler planet.

It makes for a thrilling story, putting the evolution and spread of trees at the heart of a transformation of the Devonian world, but does this chain of cause and effect stand up to scrutiny? Support for two essential links in this scenario, dubbed the 'Devonian plant hypothesis', is written in fossil soils and sediments. Fossil soils point to a changing chemical composition of the atmosphere, indicative of a massive 90% decline in carbon dioxide levels[19] that dramatically weakened the atmospheric greenhouse effect. Sediments show that the ultimate expression of this strengthening series of postulated Earth system feedbacks as a forested world took shape was the ending of the Palaeozoic greenhouse world. Out went the luxurious carbon dioxide-rich world of the early Palaeozoic. In came a new icehouse world that culminated around 300 million years ago in the Permo-Carboniferous glaciation, one of the most intense and prolonged series of severe glaciations in Earth history, with huge ice sheets extending from the poles to the tropics.[20]

On these grounds, we might argue that Earth's transition to a forested planet represents the most important natural biological feedback on climate ever witnessed over the past half billion years. We should be mindful, however, of the danger of this narrative being too simplistic, for there is a snag. It lacks the fossil evidence for early forest trees and their rooting systems engaged in weathering soil minerals in the way I have just outlined. For sure, intriguing pieces of evidence have turned up here and there[21] from fossil soils and root traces, some of them not without the whiff of controversy. As Berner would have been the first to point out, the paradigm of cause and effect outlined above largely rests on weathering by contemporary forested ecosystems extrapolated back into the past. Nothing wrong with that, you might say—we have encountered this approach to understanding the history of life and the planet before. But what about interrogating the fossil record of early trees and soils to evaluate this crucial linkage directly?

The quest for this missing link between the evolution of early forest trees and their geochemical engagement with soils and rocks brought us to a newly discovered fossil forest floor preserved in a quarry at Cairo, Greene County, New York State.[22] The site is not far from Gilboa and the rocks and sediments there also date back to 385 million years ago. Stumps of a long extinct conifer-like progymnosperm tree related to *Archaeopteris* pockmark the floor, with spidery root traces radiating outwards across the dusty surface. Alongside these are many circular depressions formed by giant cladoxylopsids. The quarry site at Cairo provided a unique opportunity—a locality with Devonian trees preserved *in situ* on the forest floor, and with tree rooting structures extending down into the sediments below that we hoped contained fossilized soil weathering profiles. Raising rock cores drilled from this ancient forest floor might, with luck, furnish vital clues for understanding how early trees and their rooting systems interacted with soils and soil minerals.

Catching a geochemical glimpse of how these ancient trees interacted with the soils of the Devonian landscape was never going to be easy. Obtaining rock cores beneath the stumps of trees dotted across the quarry floor required a mobile, tall, truck-mounted drill-rig to solve the obvious engineering challenge (**Plate 9**). The business end of the corer on the rig was a diamond-encrusted cutter, cooled and lubricated with water drawn by pump from the nearby backwaters of the Schoharie Creek. Drilling and extracting the first rock core from beneath the main lateral root system of a larger archeopterid tree stump proved a tense affair. For those of us crowding round the rig, the anxiety was palpable. As the drill-rig lowered the corer and it touched down on the thin greenish-grey siltstone capping the deeper sediments below, we had little idea what to expect. The compressed siltstone cap contains the remains of fossil fish that arrived on the fin after the forest was flooded. How deep was the layer of fossilized soil beneath the fishy siltstone cap? Would it shatter and split, destroying its geochemical story? Or would it retain its integrity and give us a shot at discovering what the roots of those early forests had been up to? As the drill penetrated the siltstone crust, coring the sediment beneath, brick-red water was flushed out of the drill-hole, painting the quarry floor bright red with 385-million-year-old ink made from fossil soils. A wave of relief and excitement broke out.[23] Standing beneath the baking sun that afternoon, we gradually realized with mounting excitement that the drilling operations were likely to succeed as more-or-less continuous sections of sediment cores were raised to the surface. The drilling campaigns that followed extracted many cores extending 3 metres beneath the major early forest trees, providing raw materials for opening up detailed lines of investigation.

The essential first step in the enterprise is to undertake painstakingly detailed sedimentary analyses down through each core, grain-by-grain, layer-by-layer, moving backwards in time as you go lower down the core. In this way, you slowly build a minutely detailed log of changes in sediment texture, such as the size and shapes of the grains, colour, and so on. Compiling these sedimentary logs for every core meant we could then align the cores obtained next to the different tree stumps across the quarry floor by comparing common features in their profiles.[24] It became clear that the stumps sat on top of a sequence of fossil soils that said something about the Devonian climate the trees experienced and the soils in which they grew. Climate clues came from careful inspection of individual grains, which showed they had what soil scientists call 'slickensided slip planes'. These distinctive features are typically only found in soils of warm sub-tropical regions experiencing seasonal rainfall, with the slickensided grains forming when soils contract on drying out and then expand on rewetting. Repeated expansion and contraction of soils causes tell-tale frictional wear of rock and mineral grains as they rub against each other. The Devonian trees, then, probably enjoyed a warm sub-tropical climate. Other features of the sediments suggested they were rooted in well-drained rather than waterlogged soils. Without this sort of evidence it is easy to become misled and Goldring herself thought the Gilboa trees had bases that were 'bulbous, as might be expected of certain trees growing under swampy conditions'. The new findings revised old views suggesting that cladoxylopsids and *Archaeopteris*-type trees grew only on waterlogged, peaty soils.[25]

More was to follow, because even casual scrutiny of the drill cores showed they displayed milky white root traces on the sides of cores (**Plate 10**). These markings, called 'drab-halos', are a record of the growth behaviour of the tree roots. A few had a central clay-rich cast running through them, preserving the structure of tree roots that once lived long ago. Mapping the depth of fossilized root traces in the rock cores showed the cladoxylopsid trees with simple roots were shallow and did not change much in depth with different tree sizes. For the conifer-like progymnosperm trees in these early forest ecosystems, the story was strikingly different, with larger trees having deeper rooting systems, and this supported a crucial missing link in the Devonian Plant Hypothesis. This sounds obvious until you remind yourself we are dealing with mysterious 385-million-year-old forest trees, and prior to this we knew next to nothing about the form and growth habit of these trees below ground.

What about evidence for the next essential link in our chain of cause and effect, deeper roots → more weathering? Clues came from the numerous root traces preserved in the fossil soils with diffuse blue-grey haloes surrounding them, indicative of chemically altered soils. Progressive changes in chemical composition down through the cores of fossil soils obtained beneath the conifer-like group showed that their roots had actively weathered rocks. Large trees caused much bigger changes in sediment chemistry than small trees. In other words the bigger the trees, the greater the intensity of tree-driven weathering.

Support for this interpretation came from clay minerals. Clays are super-fine particles made by the weathering of silicate rocks—they even occur on Mars, formed after billions of years of chemical weathering, apparently in the absence of life.[26] On Earth, though, clays form most strongly under forests because, strange as it may seem, trees are 'clay mineral factories'.[27] Trees intensify the weathering processes that erode rock into smaller and smaller particles, leading finally to clay formation. In fact, trees drive changes in the clay mineralogy of soils in a highly predictable direction, converting one suite of clay minerals into another in a characteristic manner over time, and this helps us interpret the past.[28] Since the 1960s, geologists had noted that Precambrian sediments appeared to be composed of less weathered materials, with fewer clays than younger sediments. To explain this apparent conundrum (the expectation would be that older sediments had more clays), they speculatively attributed the increased abundance of certain clays (smectite and kaolinite) in soils through the Silurian, Devonian, and Carboniferous to the actions of an expanding forested biosphere.[29] The analysis of the rock cores from Cairo suggested increased production of clays with the effects appearing to be greatest beneath the largest trees with the deepest rooting systems. When put together with the other geochemical evidence, it starts to build support for the chain of cause and effect: bigger trees → deeper roots → more weathering.

Fascinating though these studies are in helping us piece together the case for trees changing the world, they still do not yet tell the whole story. The past decade has witnessed the emergence of a new, more inclusive picture, recognizing symbiotic soil fungi as long-overlooked players in the biogeochemical cycling of elements.[30] Berner, of course, was on to this possibility straight away. He had already collected samples of Hawaiian basalt from beneath rooted plants and observed a porosity absent from the same basalt flows uncolonized by plants.[31] The explanation, he hypothesized, was 'solutions secreted by symbiotic microbiota associated with plant roots'. Under the microscope, a slide containing a cross-section of basalt rock showed a channel formed by a thread of a mycorrhizal fungus connected to a tree chemically dissolving its way through to a pocket of nutritious minerals. Had the fungus mined and supplied these to its host plant by chemically tunnelling through the rock? Others had already been thinking along these lines,[32] suggesting that the highly localized acidification occurring at the physiologically active tip of a fungal thread created a reactive front line that chemically etched out 'micro-burrows'. Alternatively, Berner's fungal thread may have simply grown through a pre-existing tunnel. Some tunnels are the work of chemistry without the intervention of biology and tiny holes and tunnels can develop as chemical reactions dissolve minerals preferentially along structural planes in the packing of the crystal lattices.[33]

What, if any, is the broader significance of mycorrhizal fungi for weathering rocks? Answering that question is crucial to the business of determining whether these microbes have been playing a hidden role in driving global change over hundreds of millions of years of Earth's history. Some insights came from field trials exploiting the superb tree collections in the Westonbirt National Arboretum, Gloucestershire in the UK. Situated in a historic landscape that has survived since Victorian times thanks to careful management by the Forestry Commission, Westonbirt is a living collection of over 3000 different trees.[34] When combined with a management team agreeable to our undertaking scientific field trials, the site provided us with the perfect opportunity to investigate rock grain weathering beneath trees whose roots are bound in tight symbiotic alliance with mycorrhizal fungi. The simple trick of incubating uniform-sized grains of silicate rocks in mesh bags buried in soils beneath the trees proved key. By carefully selecting the size of the mesh, fine tree roots were excluded but fungal threads could enter, colonize, and weather the test rock grains. In other words, incubating the rock grains in this way allowed us to isolate the actions of mycorrhizal fungi.

The secret world of microbial weathering beneath the Westonbirt trees soon revealed itself. The fungal partners of the gymnosperm trees *Metasequoia* and *Sequoia*, selected as the closest modern analogues to Devonian forest trees, proliferated in mesh bags containing basalt grains but not in those containing other less easily weatherable grains of granite or quartz.[35] Such targeted fungal proliferation reflects a tightly regulated self-reinforcing or positive feedback mechanism operating between the two groups of organisms involved—trees and fungi. As the fungi colonize the basalt grains and pass the weathered nutritious elements back to their host tree roots, the tree in return ups its provision of photosynthate to the fungus, enabling it to proliferate, do more weathering, and capture more nutrients. Now consider the reverse situation: how the feedback loop is shut down if fungi colonized unreactive rock grains. A lack of nutritious elements to pass back to the host tree means no extra carbon reward provided by the roots, slowing fungal growth. Mineral flakes recovered from the mesh bags also confirmed that fungi really do seem to be tiny trench diggers. The size and shape of the trenches measured on the flakes perfectly matched those of fungal hyphae as they etched their way across the mineral surface.[36]

With mycorrhizal fungi becoming recognized as possible agents of weathering, the focus turned to explaining how they achieve this feat. The answer comes from thinking about the flow of carbon energy between the plants and fungi. The energy flux is really a flow of sugars synthesized by photosynthesis that starts life in the tree canopy, from where it moves downwards into roots to support the pervasive growth of fungal networks in the soils below. Rock-eating microbes, it turns out, have an appetite for weathering that varies in proportion to the carbon-energy flux they receive from host trees.[37]

But how is it that cotton-wool-like fungal filaments can dissolve rock and mineral grains? The answer is through a variety of metabolic tricks, one of which involves bonding themselves tightly to mineral grains by special sticky proteins. This enables them to generate massive internal pressures—typically 4000–10 000 times higher than atmospheric pressure at sea-level[38]—to 'inflate' the fungal filament, and drive their exploratory growth. Such pressures may even be an evolutionary adaptation for penetrating rock surfaces and, in the case of pathogenic fungi, plant tissues. By bonding tightly to the surface of the mineral, the fungus can generate spectacular downward forces that mechanically fracture rock grains. For good measure, they combine this mechanical pressure with an acid bath treatment, thanks to the extremely high metabolic rates of the hyphae, generating carbon dioxide, the waste product of aerobic respiration, which acidifies the aqueous soil solution in which they live. Acidification is highly localized by rapidly proliferating hyphae. Exuding streams of hydrogen ions during nutrient uptake, they weaken the already pressured mineral surface, causing it to split and fracture. Every split and fracture opens up new faces for the chemical weathering reactions to proceed, hastening destruction. Countless millions of hyphae are, unnoticed, eating their way through countless rock grains in soils of the world, supported by carbon supplies delivered from the trees above ground.[39]

Bringing mycorrhizal fungi into our picture of events in the Devonian, adds a new link to the chain of cause and effect involving the soil nutrient, phosphorus.[40] Phosphorus is an essential element for all forms of life, which need it for making DNA and cell membranes, amongst other things. In the early stages of soil formation, it is provided by the weathering of the mineral apatite because most of the phosphorus in the Earth's crust (over 95%) is locked up in apatite, with far smaller amounts occurring as inclusions in igneous rocks. On land, apatite stocks are being slowly but progressively depleted by weathering. Stocks are only recycled when plate tectonics uplifts buried rocks containing apatite or when volcanic eruptions at plate boundaries spew out huge amounts of fast-weathering basalt. Apatite depletion on land means that, in the long run, the productivity and biomass of the trees declines as they exhaust rock phosphorus supplies.[41] Consider how this might have played out with the arrival of trees and then forests of trees in the Devonian. As early trees and forests evolved into the major forms, including those uncovered by Berry, Stein, and colleagues, what better way to meet their growing demand for phosphorus than by teaming up with mycorrhizal fungi that are highly adept at mining it from apatite in soils? It is easy to imagine these early forests expending carbon energy to support mycorrhizal fungal partners in return for obtaining access to the phosphorus supplies necessary for running their metabolism and building biomass.

At this point, however, we should acknowledge that evidence for fossil mycorrhizal partnerships in the roots of trees forming Devonian forests is lacking. We should be mindful of the limited nature of the fossil record of these trees and its generally poor quality of preservation. Recall that forests like those in Gilboa have little of the original anatomy remaining. Here, the dictum 'absence of evidence is not the same as evidence of absence' is apposite. By the time of the Carboniferous, arborescent lycopsids, which dominated the extensive tropical swamplands and rose to heights of over 30 metres, were mycorrhizal, as we know from well-preserved fossilized strap-like root appendages preserving arbuscules and other diagnostic features of the fungi. These giants of the past have since become extinct but we also know that all of the extant gymnosperm groups that originated at least 300 million years ago (including cycads, ginkgo, and others) form mycorrhizal partnerships that facilitate the release of phosphorus from apatite.[42] There is also the constancy of mycorrhizal alliances of trees since those far-off days of the Carboniferous because all later-evolving tree groups have roots tightly bound into associations with symbiotic fungi. So it is simply inconceivable to imagine Devonian trees as being any different to modern trees in terms of relying on fungal symbionts to fulfil their nutrient demands as they evolved progressively larger stature.

A revised hypothesis explaining how the rise of Devonian forests bioengineered Earth's climate as they 'greened' the continental land surface might, then, look something like this. As trees with bigger and architecturally more complex rooting systems evolved, they pumped more carbon energy below ground, fuelling the rock mining by roots and mycorrhiza for phosphorus and other essential elements needed to meet the rising demands for synthesizing more tissues. These activities, in turn, acidified soils and increased the biologically driven weathering of rocks. The carbon fluxes involved were likely enormous. Forests and grasslands of the modern world direct a carbon-energy flux estimated to be equivalent to around six times our annual electricity production from burning fossil fuels.[43] Or, put another way, three to seven orders of magnitude greater than the kinetic energy generated by tectonic uplift traditionally thought to regulate the global biogeochemical cycling of elements over time.[44]

Viewed in the light of such mind-boggling numbers, it starts to become clearer how the arrival and spread of forests through the Devonian might have bioengineered global climate, ramping up the carbon-energy flux pumped into soils via allocation to roots and mycorrhiza to influence rock weathering and atmospheric carbon dioxide sequestration.[45] We can capture nutrient effects simply by adding an extra step in our logical sequence. Nature abhors destabilizing or self-reinforcing (positive) feedback loops and yet by doing this we find one that looks like this: bigger trees → deeper roots → faster weathering → greater nutrient release → bigger trees.

## PEAK PHOSPHORUS?

Humanity, like the emerging forests that 'greened' the continental land surfaces, has a huge and growing demand for phosphorus. We rely on rock phosphate that formed 10–15 million years ago for supplies to make fertilizers that support modern agricultural production. Unlike nitrogen fertilizers synthesized industrially with the Haber–Bosch process, we are unable to synthesize phosphorus in the lab. Like time itself, phosphorus is non-renewable. There is no substitute.

Yet excess phosphorus is being wasted as it flows from fertilizers to watercourses and coastal oceans, fuelling intense algal blooms that deliver pulses of organic matter to sediments. Microbial decomposers consume dissolved oxygen, creating gigantic 'dead zones' in the coastal oceans—affecting over 250 000 $km^2$ in 2008.[46] The anthropogenic black-shales of the modern world that will eventually form around the coasts will tell our story to future generations of geologists reading humanity's footprint on the sediments of the planet.

Meeting the agricultural demand for fertilizers saw global extraction of phosphate rock triple since the end of the Second World War.[47] Costs rise as the quality of phosphate rock reserves declines, raising the question: are we approaching peak phosphorus production, a time when we are consuming it faster than we can economically extract it? Peak production might be on the horizon within a few generations, but an accurate timeline depends on assessing the quantity remaining in the ground, and the extent we recycle. Gauging reserves is dependent on voluntary provision of data and old geological assessments.[48] Currently, the four top countries thought to control more than 85% of known phosphorus reserves, are Morocco, China, Algeria, and Syria, with Morocco in the driving seat. By contrast, a dozen members of OPEC control 80% of the world's oil reserves.

There is no 'quick fix'. We must face up to the prospect of recycling phosphorus from human waste, as the French poet and author Victor Hugo (1802–1885) highlighted in *Les Misérables*:

> Science, after having long groped about, now knows that the most fecundating and the most efficacious of fertilizers is human manure. The Chinese, let us confess to our shame, knew about it before us... If our gold is manure, our manure on the other hand, is gold.

Toxins, drugs, and heavy metals rule out the use of untreated sewage, but the idea is getting a modern resurrection, with companies converting ashes from the combustion of sewage sludge in incinerators into high-quality, phosphate-rich fertilizer fit for crops.[49] Adapting agricultural management practices to reduce phosphorus demand is also needed.[50] In 1938, US President Franklin Roosevelt said it was 'high time for the Nation to adopt a national policy for the production of phosphates for the benefit of this and coming generations'.[51] Seventy-five years later, we are still searching for long-term solutions to the phosphorus problem.

The question raised by uncovering potentially destabilizing climate feedbacks in explaining the role of trees and forests in air conditioning the planet through the Devonian is this: why has Earth not experienced runaway cooling since they appeared on the continents? A system of checks and balances has to be operational in stabilizing changes in atmospheric carbon dioxide levels to have prevented Earth's climate spiralling out of control. Geochemists believe that these stabilizing (or negative) feedback loops constitute a planetary thermostat, involving weathering, carbon dioxide, and climate.[52]

To imagine how the thermostat might operate, consider this scenario. Suppose atmospheric carbon dioxide levels rose due to a massive and prolonged episode of volcanic activity. This would drive a warmer climate, due to an enhanced greenhouse effect, and a wetter climate, by causing a more vigorous precipitation cycle. Both features accelerate chemical reaction rates of rock weathering, which remove carbon dioxide from the atmosphere and cool the planet. The thermostat also operates in reverse if atmospheric carbon dioxide levels are falling, thereby slowing weathering and allowing carbon dioxide to build up. Suppose, for example, the uplift of major mountain range exposed large amounts of fresh unweathered rocks to the atmosphere. These rocks can now react with mildly acidic rainwater to remove carbon dioxide from the atmosphere as they undergo weathering. The rise of the Himalayas is a good example of such a tectonic event. It is linked causally to a global decline in the atmospheric carbon dioxide concentration and climate cooling, as witnessed by the ensuing glaciation of Antarctica, and a slower weathering regime.[53]

Of course, there are more processes involved, and for rock weathering to operate as a climate feedback the Earth must be sufficiently tectonically active to ensure an adequate supply of fresh rocks and minerals. In this context, the dramatic 90% drop in atmospheric carbon dioxide caused by the rise of trees through the Devonian is analogous to the uplift of a massive mountain range. It switched Earth into a mode of accelerated weathering, sequestering atmospheric carbon dioxide, and cooling the climate to resist further weathering. Indeed, there was also the uplift of an equatorial mountain range in the Permo-Carboniferous that may have reinforced the effects of trees in cooling the planet. Conversely, its later destruction may also have contributed to putting the brake on weathering, checking further cooling.[54]

Shifts in the balance between volcanoes adding carbon dioxide to the atmosphere, and rock weathering slowly removing it, are reflected in Earth's atmospheric carbon dioxide history, as reconstructed from a variety of fossil materials for

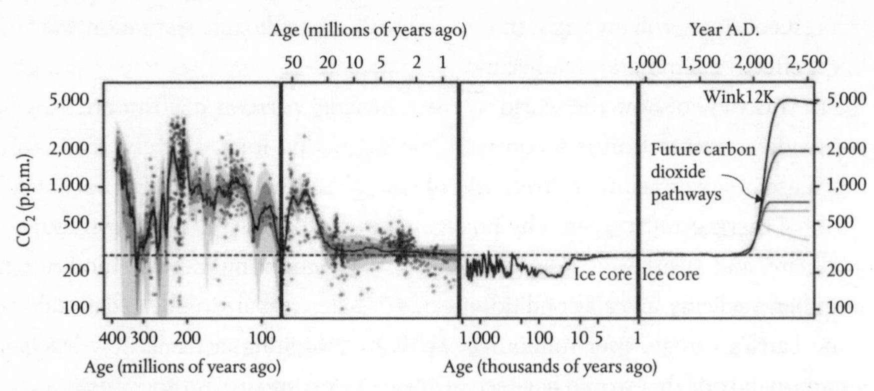


**Figure 24** Earth's atmospheric carbon dioxide ($CO_2$) history over the past 420 million years plotted alongside possible future pathways of increases in comparison. Notice the apparent minimum threshold or lower limit that has prevailed over the past 20 million years. Dashed line indicates the current average $CO_2$ concentration (405 p.p.m. in 2017)

the last half billion years (**Figure 24**).[55] Careful inspection of this impressive reconstruction reveals a curious feature—when the atmospheric carbon dioxide concentration falls, it seems to bump up against the same minimum value, over and over again.[56] Over the past 24 million years, in particular, carbon dioxide levels have flat-lined, even, surprisingly, during major mountain-building episodes like the uplift of the Southern Alps in New Zealand and the Andes. According to the geochemists, the production of huge quantities of fresh rock during these tectonic upheavals should have enhanced weathering and sucked carbon dioxide out of the atmosphere. Yet for some reason, it stayed put. If the reconstruction is valid, the pattern suggests the Earth system may have a built-in minimum atmospheric carbon dioxide concentration. The question is: do the mechanisms of this strange phenomenon extend beyond the imaginations of hard-rock geochemists?

To be sure, the identity of the mysterious 'doorstop' limiting the minimum amount of carbon dioxide in the atmosphere is uncertain—is it some inherent property of Earth's natural thermostat? Another line of enquiry suggests that trees and forests could be involved.[57] The basic idea is that as the global atmospheric carbon dioxide concentration begins to dwindle during major episodes of mountain uplift, it gradually starts starving the trees and forests colonizing these uplifted terrains. As it dips towards 200 parts per million, half what it is today, carbon dioxide starvation compromises the health of the world's forests, causing tree productivity to crash.[58] Dead trees are unable to weather rocks actively, and forests of slow-growing trees, thinned out by carbon dioxide starvation, become less effective at mineral weathering.[59]

If this view of how the world works is broadly correct, atmospheric carbon dioxide concentration is a control knob regulating forest-driven weathering, just as it regulates climate. Trees are integral components of the natural built-in global thermostat regulated by linkages between atmospheric carbon dioxide, climate, and weathering. Our debt to plants is mounting. Besides food, water, timber, medicine, and air-conditioning our climate, trees are implicated in stabilizing Earth's climate over millions of years by inhibiting critically low levels of carbon dioxide that would have seen Earth develop into a giant snow ball.

We can see from this that ultimately rock weathering will remove our anthropogenic carbon dioxide emissions from the atmosphere and eventually deposit it on the seafloor as carbonate sediments. Unfortunately, in normal circumstances rock weathering is far too slow to help avert the threat of dangerous climate change. The basaltic terrain of Iceland is already responding to dramatic warming of the Arctic region over past decades by weathering faster,[60] but this feedback will not save us. Our best estimates suggest natural rock weathering sequesters less than 2% of our annual carbon dioxide emissions from fossil fuels. At current rates, it will take a hundred thousand years or more for natural weathering processes to convert our accumulated emissions of carbon dioxide into anthropogenic carbonate deposits on the seafloor. Yet we find ourselves in far from normal circumstances, with an atmospheric carbon dioxide concentration exceeding 400 parts per million for the first time in several million years, thanks to our combustion of fossil fuels. In 2016, our annual emissions of carbon dioxide reached 36 billion tonnes, the highest in human history and 60% higher than in 1990.[61]

Earth's atmospheric carbon dioxide history can be compared with what might happen in the future. The chart in **Figure 24** show potential pathways we may follow, depending on how much fossil fuels are burned over the coming decades. Steep changes correspond with rapid future climate change. If we keep burning fossil fuels, we could soon cause higher carbon dioxide levels and faster climate change than the Earth has seen in the past 50 million years. If we burn all available fossil fuel reserves (the black 'Wink12k' line), the carbon dioxide concentration will exceed levels seen in the entire 420-million-year reconstruction.

The continued accumulation of carbon dioxide and other human-caused greenhouse gases in the atmosphere since the pre-industrial era (i.e., from around

1850 onwards) has already driven global warming exceeding 1°C above the pre-industrial value.[62] In 2016, the rise in annual global temperature was almost 1.3°C. If warming continues at the current rate, the aspirational target of the recent United Nations Paris Agreement[63] of 1.5°C will be out of reach within 20 years, threatening a fifth of all species with extinction. Breaking our addiction to fossil fuels and phasing over to carbon-free energy will not be easy to achieve, and makes even a more lenient target of 2°C difficult. Urgently phasing down carbon dioxide emissions is the *sine qua non* for fighting the threat of future climate change. There is also a growing realization that research into safe and affordable methods for extracting carbon dioxide from the air is required to augment efforts to reduce fossil fuel emissions. All such strategies are currently poorly understood, especially in terms of their environmental and ecological impacts, financial costs, and feasibility.[64] But is it possible for humanity to mimic Devonian trees in converting huge amounts of atmospheric carbon dioxide into carbonate minerals to cool the planet?

One approach proposes to deals with point sources of carbon dioxide in this way from power stations. Approximately 2200 coal-fired power plants in Europe, North America, and China release about a third of fossil fuel carbon dioxide emitted into the atmosphere.[65] If we could retrofit these coal-fired plants with carbon capture and sequestration technologies, it would slow emissions into the atmosphere—assuming we could find somewhere to store it and the economics work. Perhaps we might bury the carbon dioxide by reacting it with basalt directly as the Earth has been doing for billions of years, aided in the past few hundred million years by trees? Basaltic rocks represent huge underground storage depots that are sufficiently large to accommodate our carbon dioxide emissions for centuries to come. Once again, researchers have turned to Iceland where, like Berner, they seek to exploit its remarkable basalt landscape.

Located 25 kilometres east of Reykjavik, the CarbFix experiment in Iceland is a sequestration project assessing the potential for permanently capturing and storing carbon dioxide gas as carbonate minerals in basalt rock.[66] In 2002, CarbFix injected hundreds of tonnes of carbon dioxide dissolved in groundwater into basaltic rocks to a depth of nearly a kilometre. Within a couple of years, over 95% of the carbon dioxide reacted with the basalt to form environmentally benign carbonate minerals.[67] Carbonate minerals are stable residents of geological strata, stable enough to lock up carbon dioxide for millions of years. The speed of conversion of carbon dioxide to carbonates has proved surprisingly quick. The common view was that immobilization of carbon dioxide in this way would take hundreds of years, if not longer. Effectively, this small-scale programme has demonstrated the initial feasibility of safely storing anthropogenic carbon dioxide emissions below ground in basaltic rocks. Further support comes from another pilot project, this time in the north-western United States, where they have already injected 1000 tonnes of pure carbon dioxide into the Columbia River basalt formation near the town of Wallula.[68] How it might play out in the longer term is unknown. Could the porous basalt become plugged with carbonate minerals near the injection site, blocking the spread and capture of carbon dioxide? Encouraging though these findings are, a safe solution for permanent carbon dioxide sequestration in this way still seems a long way off. Even if technical hurdles could be overcome, the financial hurdles are, at present, impossibly high.

Many understandably question the enormous scale of investment in infrastructure required to extract massive amounts of carbon dioxide from the atmosphere and the burden this would place on future generations.[69] Others raise concerns over the large tracts of land that might need to be set aside for planting and harvesting bioenergy crops by the second half of the twenty-first century as a potential alternative to fossil fuels.[70] And what about the water demands and energy costs involved in making fertilizers to grow the energy crops? Is there a better way? A few years ago, a team of us proposed another approach, this time adopting managed croplands for carbon capture and storage by utilizing them to weather silicate rocks and, in particular, basalt.[71] Amending soils of managed croplands with crushed fast-reacting silicate rocks could accelerate their chemical breakdown, pulling carbon dioxide from the air into soils and, eventually, the oceans, to cool the planet, mimicking the bioengineering actions of trees in the Devonian. These processes also release nutrients that fertilize crop growth. And, as crops take up dissolved silica released from weathered rocks, they build tougher cells and prime their immune systems, giving protection against herbivore pests and microbial diseases. Such benefits could lower agricultural fertilizer and pesticide usage and costs to improve the profit margins of farmers.[72]

With nearly 11% of the terrestrial surface annually managed for crop production, this could offer an opportunity to deploy a means of carbon sequestration on a large scale within a decade or two, and also improve food security. Rapid deployment is important if we are to use technologies to significantly draw down carbon emissions by the latter part of this century. Undertaken carefully to avoid undesirable consequences, a well-designed programme might also restore soils, improve crop yields, and conserve geological fertilizer resources, especially rock phosphate. It could also benefit the marine environment by lowering carbon dioxide levels and generating soluble alkalinity to help counter ocean acidification, which threatens coral reefs and marine fisheries. Ocean acidification occurs when carbon dioxide dissolves in the seawater to form carbonic acid, lowering pH levels and making the water more acidic. Over the past thirty years, the ocean's acidity has jumped by 30% and continued acidification may cost the global economy, including loss of fisheries harvest, up to $1 trillion a year by 2100.[73]

Would this ambitious and innovative scheme be feasible, or even desirable? On the feasibility question, arable farming is already equipped for application of lime and fertilizer granules, making annual applications of ground basalt at scale possible. Application rates would likely have to exceed those used in typical liming operations, but the demand for reactive silicate rocks if rolled out globally would be huge. Recycling massive quantities of freshly produced plant nutrient-containing artificial silicates[74] could help meet the demand; exploiting legacy reserves that have accumulated over time would help further. These materials include basic iron and steel slag, which have a long history of farm use in place of lime.[75] Increased construction and building activities in Brazil have promoted the exploitation of basaltic reserves, and interest is growing in recycling accumulating fine basalt dust waste as an agricultural fertilizer. If necessary, mining wastes could be supplemented with substantial silicate rock resources. There is no lack of basalt formations on land. Basalt is one of the most common rock types on Earth, covering around 10% of the land surface.

Like other potential large-scale strategies for removing atmospheric carbon dioxide, enhanced weathering on land requires further research, development, and demonstration. If proven effective, and environmentally sound, significant potential exists for meaningful large-scale deployment to capture atmospheric carbon dioxide. Extensive deployment of any atmospheric carbon dioxide removal approach has to be socially and environmentally acceptable and requires risk assessment, public participation, and transparency. To be clear, though, this will not solve the problem of climate change. Drastic near-term emission reductions are required and delaying them makes the problem progressively harder to solve.

Taking control of the global biogeochemical cycles that have shaped Earth's history to clean up our atmospheric carbon pollution and cool the planet is a formidable and unwanted challenge. Yet continued emissions will force society to consider expensive, arguably implausible, industrial-scale carbon clean-up operations to stabilize climate, with unknown ecological, environmental, and social consequences. Or face growing and alarming climate impacts including intensifying droughts, heat-waves, and storms. Trees began bioengineering planetary climate millions of years ago in the Devonian, but back then they had time on their hands. Our current crisis is urgent and unfolding at a time when global food demand will more than double before the end of the century.[76] Can we sustainably feed a crowded planet, preserve the wonderful diversity of life on Earth, and stabilize the climate? These are the pressing challenges facing humanity that we consider in Chapter Eight.

# 8 EDEN UNDER SIEGE

'Man is everywhere a disturbing agent. Wherever he plants his foot, the harmonies of nature are turned to discord.'

George Perkins Marsh, *Man and Nature*, 1864

Peter H. Raven delivered his case in a succession of quick-fire staccato sentences, each a condensed nugget of startling information giving you pause for thought. He maintained his rapid-fire delivery for an hour and received a standing ovation for his address from an appreciative, if stunned, audience. His take-home message hit you straight between the eyes: the world of plants is beautiful and fascinating, but it is facing unprecedented threats. In fact, all of nature is in serious trouble. Worldwide, Earth's biodiversity is declining at alarming rates and showing no sign of slowing.[1] Vulnerable plants and animals are fleeing and disappearing under the continent-wide shadow of humanity. Raven, Director of the Missouri Botanical Garden[2] for forty years, and world-leading advocate of biodiversity conservation, was speaking in Melbourne in 2011, at the 11th International Botanical Congress, one of the eagerly anticipated meetings of thousands of botanists from all over the world that takes place every six years.[3]

Put bluntly, the grim state of nature is our fault. Humans are a ridiculously successful species of primate, with our population growth tied to increasing agricultural output for over 10 000 years.[4] Between 1900 and 2000, the world population grew from 1.7 billion to 6 billion, supported by a six-fold increase in agricultural production made possible by nitrogen fertilizers synthesized by the Haber–Bosch process, alongside other advances in science and medicine.[5] Nitrogen fertilizers nourish the crops, and the animals fed on crops, which we consume to give us nine-tenths of the essential amino acids needed to build proteins. This holds true in developed and developing nations of the world including China, Egypt, and Indonesia. As the Czech-Canadian scientist and policy analyst Vaclav Smil reminds us, 'when you travel to Hunan or Jiangsu, through the Nile Delta, or the manicured landscapes of Java, remember that the children running around or leading water buffalo got their body proteins via the urea their parents spread on fields, from the Haber–Bosch synthesis of ammonia.'

World population is currently 7.6 billion, having increased by some 80 million annually for the past 40 years.[6] According to United Nations forecasts,[7] world population will increase to over 9 billion by 2050 and a little over 12 billion by 2100. Analysts suggest there is little prospect of an end to world population growth this century without unprecedented declines in fertility throughout sub-Saharan Africa, a region experiencing fast population growth. If the population of Africa grows by billions, it could lead to severe resource shortages, and possibly reductions in population size through mortality or fertility effects. Globally, feeding a world population of 12 billion by 2100 brings with it daunting challenges in terms of food, water, and energy security, and an escalating threat to biodiversity.[8]

Continued growth in the human population and our unquenchable demands for sybaritic lifestyles are driving unsustainable exploitation of the Earth's natural resources. As our per capita demand rises, we degrade the vital ecosystem processes that constitute our planet's life support system.[9] No aspect of the Earth is untouched by human actions. Pressure to feed the planet by harvesting the biosphere has seen nearly 5 billion hectares of the Earth's land surface converted to agricultural land.[10] Dendritic transport networks dissect and fragment even more of the landscape, and ~43% of all land is now under the direct influence of humans—30% of wilderness areas in the Amazon have been lost and 14% in Africa.[11] Indeed, deforestation, over-hunting, and invasive species are still the dominant drivers of species loss.[12]

But alongside deforestation brought about by the continued expansion of agricultural land into pristine forests emerges the threat of climate change resulting from exploitation of fossil fuels to meet rising energy demands. Combustion of fossil fuels for energy releases 36 billion tonnes of carbon dioxide into the atmosphere annually,[13] and has raised the concentration of carbon dioxide above 400 parts per million for the first time in human history, and probably in over 20 million years.[14] Already, the accumulation of carbon dioxide and other human-caused greenhouse gases is warming the planet, melting ice sheets, glaciers, and permafrost, and triggering the poleward migration of species.[15] Welcome to the Anthropocene[16]—a geological epoch defined by the pervasive and persistent signatures left in ice, sediments, and rocks by the activities of humans.[17]

The pressures on plants and animals as humans consume an ever-larger share of the planetary pie by putting land to the plough are immense. In 2016, the Royal Botanic Gardens, Kew, provided the first global assessment of our floras in its *State of the World's Plants* report.[18] Kew scientists estimate that over 20% of the world's vascular plant species are at risk of extinction, with a further 4% critically endangered.[19] From an estimated pool of over 450 000 species of plants worldwide,[20] this means 90 000+ species are at risk. A further 10% of species in the planet's plant portfolio are classified as 'near threatened', which means that without conservation actions they may also face extinction. Over 30% of all cacti, relatively recent additions to our planet's floras, are facing threats from unscrupulous collectors of live plants and seeds.[21] Plants, now more threatened than birds, are facing a similar level of threat to mammals. The biggest threat to plant species is the destruction of habitats caused by deforestation, usually for farming and cattle ranching, closely followed by selective logging for valuable timber. A major problem is that vulnerable species exhibit highly clumped distributions over the world's land surface. This creates regions of exceptionally high floristic diversity—biodiversity 'hotspots'. Most of these hotspots are located in the tropics where the human population is expanding rapidly, destroying and fragmenting habitats.[22]

**Figure 25** Botanist Hewett Cotrell Watson (1804–1881).

## HEWETT WATSON AND ESTIMATING SPECIES EXTINCTION

In 1859, the same year as the publication of Darwin's *Origin of Species,* English botanist Hewett C. Watson (1804–1881) established the mathematical basis for converting habitat loss into species loss (**Figure 25**). Born in the South Yorkshire village of Firbeck,[23] Watson was brilliant intellectually and cantankerous. He generously provided Darwin with detailed scientific information on British plants, and once even turned down the opportunity to meet the great man, claiming he was too busy. Nevertheless, the two men remained friends, and with the publication of *The Origin of Species,* Watson was among the first to write Darwin a congratulatory note on his extraordinary achievement.

Watson's knowledge of British vascular plant floras was unequalled, and he laid the foundations for one of ecology's few laws: the species–area relationship.[24] He showed that if you plot the number of vascular plant species in successively larger sampling areas, ranging from his small garden plot in Surrey, which was only a few square metres, up to that of all England, the number of species increased logarithmically. We now know that this pattern holds for geographical areas larger than England, with the total number of species increasing at a proportionally faster rate. Ecologists describe such a relationship with a power-law equation that underpins the theory of island biogeography. In its modern form, Watson's species–area relationship offers a basis for predicting how many species should become extinct as the size of the 'island' of forest in a 'sea' of cleared land shrinks.

How many species of plants are likely to be lost through continued habitat destruction? How long will it take for species extinctions to unfold? Converting habitat loss into species loss involves combining a modern form of Watson's species–area relationship, which provides an estimate of the number of species in a given area, with projected rates of habitat destruction. From this, you can estimate how many species may go extinct as the islands of forest shrink. These analyses show that because of the non-linear nature of the relationship, habitat loss initially causes little extinction, with the extinctions ramping up as the last remnants are destroyed. Assuming the rate of forest destruction stays the same, the crude extinction curve peaks in 2050 with an estimated 50 000 species lost per decade. However, the situation is worse than this analysis implies, because it assumes species are evenly distributed throughout the forests. In reality, this is not the case, many species of plants and animals exhibit highly clumped geographical distributions, often with small ranges. As a result, 44% of vascular plant species are confined to just 25 'hotspot' regions that comprise no more than 2% of the Earth's land surface (**Figure 26**).[25] Of these, 17 are in tropical forests, which have already suffered a disproportionate loss of primary habitat, threatening many species with extinction. Consequently, habitat destruction acts like a 'cookie cutter' pressed into poorly mixed dough.[26] Those species with small ranges in the stamped-out area are eliminated, while those which are more widely distributed survive. If we protect all the remaining habitats in individual hotspots, our species–area curve predicts that 18% of their species will eventually become extinct by 2100. If the higher-than-average rate of habitat loss in these areas continues for another decade, then around 40% of species could be lost by 2100. Immediate action to protect the remaining unprotected habitats of the humid tropical forest biodiversity hotspots could drastically cut the numbers of species that will eventually be lost.

On top of competition for land to meet our rising demands for food and natural resources sits the imminent threat of human-made global climate change to species numbers. Estimates of species loss caused by climate change also often hinge on Watson's species–area relationships,[27] with the area defined by the boundaries of the climatic conditions a species of plant or animal encounters throughout its geographical range. Ecologists call this the 'climate envelope' of a species. They represent 'islands' of acceptable climate in a 'sea' of

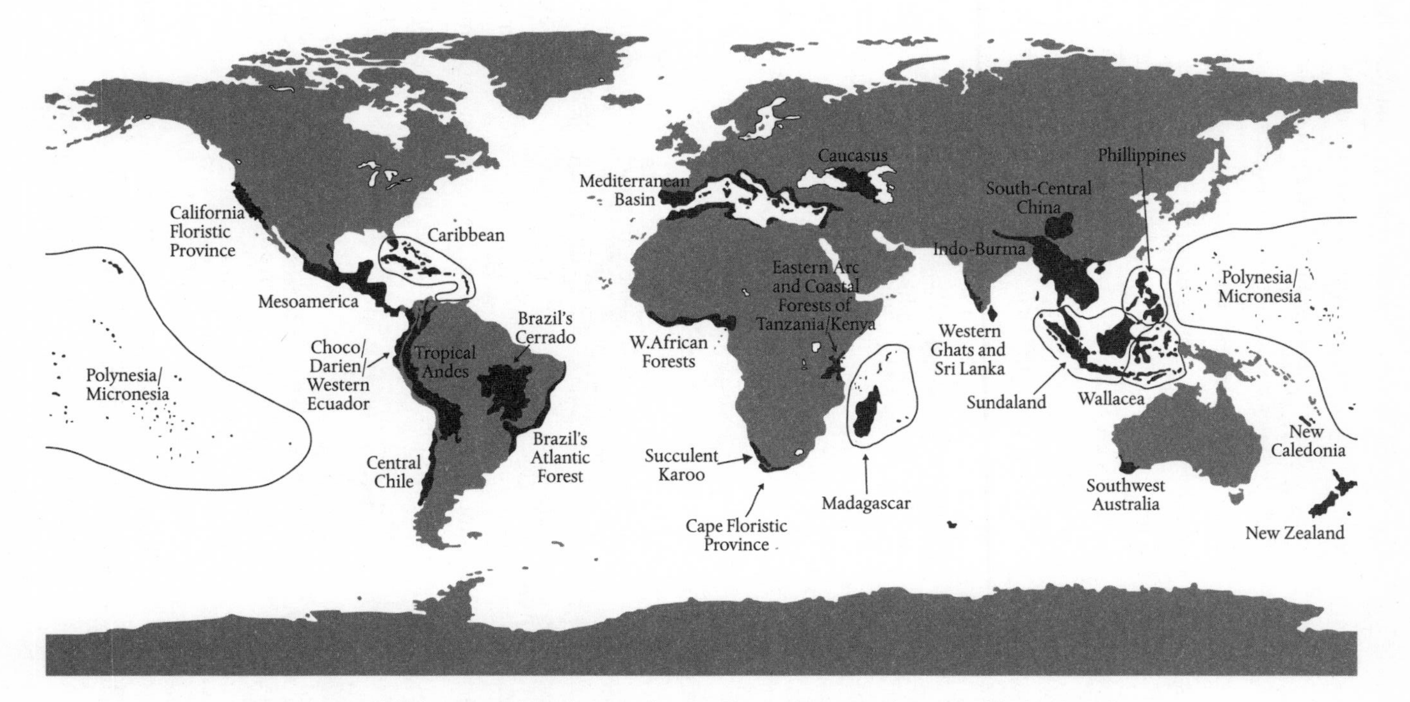


**Figure 26** Global terrestrial biodiversity hotspots cover no more than 2% of the land surface.

## AMAZONIAN BIODIVERSITY UNDER THREAT

The largest remaining block of the tropical moist forests that once covered 15% of the land surface is the Amazon, where deforestation, logging, and human-caused fires symbolize the destructive forces of humanity. Unsustainable logging and burning is degrading and destroying the complex habitats, while road building and mining is fragmenting much of what remains, especially in the Brazilian Amazon. This extraordinary biodiversity hotspot contains an estimated 11 000 tree species.[28] The good news is that over 3000 of these species have sufficiently large populations of individual trees (at least a million) to ensure they are likely to persist into the future. The bad news is that up to half of the 5000 or so rare Amazonian tree species (i.e., those with less than 10 000 individuals) face extinction. Such small populations of rare tree species with limited genetic diversity are highly vulnerable to extinction by local habitat loss and climate change. These assessments are uncertain and hampered by incomplete knowledge of the biogeography, life history, and environmental requirements of tropical trees. Here's the point though. If the trees go extinct, the complex ecological network of organisms that make up a once pristine tropical ecosystem collapses along the pathway of diminished diversity: trees → vertebrates → invertebrates → microbes. So too if the animals that play a crucial role in pollination, and seed and fruit dispersal, go extinct, but the direction of causality in the 'silent forest' is reversed.[29] In Amazonia, some ecologists estimate that 70–80% of extinctions of forest-dwelling vertebrates caused by rainforest losses are yet to come.[30] On the plus side, this time delay is also an opportunity for restoring habitats or implementing alternative measures to safeguard species otherwise committed to extinction.

unsuitable climate prevailing across the rest of the globe. Using this information, you can calculate how the geographical envelope of acceptable climate for different species changes as the climate warms in response to rising atmospheric greenhouse gas concentrations. If the potential geographical range of a species shrinks, its risk of extinction increases. The approach is irresistible to anyone with access to databases of species distributions, the outputs of climate models predicting future climate change, and modest programming skills;[31] forecasting patterns of extinction in this way has become a cottage industry. Although many assumptions (e.g., are the soils suitable? what are the effects of competition? etc.) lie behind these beguiling projections, the basic underlying idea is probably correct—less habitat, fewer species. What these models hint at, then, despite the assumptions, is the size and locations of extinctions that future warming might cause.

Extensive analyses assessing the threat to over a thousand species of plants and animals suggest that a warming of 0.8–1.7°C by 2050 commits nearly a fifth of species to extinction. This startling figure rises to a quarter of species with 1.8–2.0°C warming, and over a third with >2.0°C of warming. For plants, 20–50% of endemic species could be committed to extinction over the next 50 years. Endemics are highly prized by conservation biologists because they are unique to a specific region, making their preservation a priority. In projections from this work, warming drives the geographical ranges of many species polewards and towards higher elevations on mountains.

So much for global analyses, what about the European scene? In Europe, assessment of the climate change threat to plant diversity is flagged as severe, with over half of the 1500 species investigated threatened with extinction by 2080. Species living in mountainous regions of the Northern Mediterranean experience the greatest projected declines, with up to 80% species loss,[32] and it is easy to understand why. Mountain plant species are specialists in coping with the harsh conditions. Over time, they have evolved adaptations for exploiting the limited opportunities to grow and reproduce under montane conditions. Faced with a future warming climate sweeping across the Alps, Pyrenees, central Spain, French Cevennes, Balkans, and Carpathians, the narrow cold-climate tolerances of mountain floras means the only way for species to cool off is to gain elevation. Many species in alpine and subalpine regions of Europe and North America are already responding in this way, migrating to higher elevations to escape the past century of climate warming.[33] Indeed, analysis of observations and measurements on over 300 European mountains reaching back over a century and half suggests mountain biodiversity is increasing as warming accelerates and species migrate upwards and gather towards the summit.[34] Does this foreshadow the loss of species in the near future, as they become pushed off the face of the planet when they run out of mountain? Perhaps in southern Mediterranean regions, plants may fare better because the species there are adapted to the hot, dry, summers, tolerating heat and drought, meaning they may cope better with future climates.

As we can see, some of these alarming predictions are already playing out as the theorists suggest, but it would be a mistake to accept them unquestioningly at face value.[35] Could, for example, rising atmospheric carbon dioxide concentrations partially ameliorate anticipated ecological damage caused by warming? If the carbon dioxide-rich world expected for 2050 fertilizes plant growth, it might increase the capacity of some regions to support more species.[36] The idea is that diverse assemblages of plant species are generally more productive, and higher productivity might generate ecological elbow room for more species. Such is the extent of human intervention in eroding biogeographical barriers and modifying species distributions by moving and introducing plants to new places, the rapid arrival of alien species could conceivably fill the extra capacity of the environment. In other words, immigration is an important input term to consider in the biodiversity picture. In many regions of the world, the introduction of non-native species has greatly increased plant species richness, and even generated new species via genome expansion through hybridization.[37]

Probably it is too early to draw any comforting reassurances from these twists to the guessing game on species numbers. None of the pessimistic projections of doomed diversity yet account for the accelerating pace of future climate change, its velocity of assault on the natural world. Can plant and animal species reach the escape velocity they need to avoid extinction, or will the changing climate overwhelm the capacity of species to maintain themselves under the stable conditions they have experienced over the past 10 000 years? Factor velocity of climate change into the mix and we find that, over 30% of the Earth's land surface, the speed at which plant species must migrate to keep pace with a shifting climate in the coming century exceeds their capacity to disperse.[38] Those species left behind in the race to keep up with the fast-moving climate envelope may perish. Current projections indicate this is likely to be a common problem confronting plants, because climates found today on 20–50% of the planet could disappear within a century.[39]

Unfortunately, the extinction story does not end there, because our actions now are leading to an 'extinction debt' payable by the natural world at a later date.[40] Extinction debts arise because of the time lag between habitat destruction or climate change and the subsequent terminal decline of populations of plants or animals. Forecasts of extinction based on species–area curves overlook this crucial issue, assuming instead that extinctions are instantaneous, reductions in numbers following lock-step with a shrinking habitat area or contraction of an agreeable climatic envelope. But in reality plant extinction is likely to be a slow-burn phenomenon, and for good reason, as tropical ecologist Dan Janzen noticed when agriculture replaced native forest in Costa Rica. In the newly created agricultural landscape, surviving patches of forest containing native trees persisted in field margins, and a few pockets of diversity elsewhere, but were unable to reproduce because the habitat suitable for seedling establishment had been destroyed. These trees, Janzen wrote, are the 'living dead'; sitting in the extinction waiting room.[41]

A clue to the sort of timescale involved in plant extinctions comes from Napoleon's St Helena, the remote South Atlantic Island where the tree habit has evolved with alacrity. When Portuguese navigators discovered the island in 1502, they introduced goats, which, without natural predators, soon multiplied into huge herds. Ravaging goats and people destroyed the vegetation, and yet some plants like the St Helena olive (*Nesiota elliptica*), unrelated to the true olive, hung on with remarkable tenacity.[42] Despite there being fewer than 10 trees by 1900, the St Helena olive finally went extinct in 2003. Extinction time lags for woody plants on St Helena are on the order of a century or longer,[43] and this may be the rule of thumb for most plants. Individuals of long-lived plants, like trees, for example, may continue to reproduce or, like Janzen's trees, simply live on without reproducing, even if climate change or habitat destruction means viable populations are unsustainable over time. Plants can also regenerate from seed dormant in the soil (until the numbers of seeds are exhausted) and can reproduce by making clones, meaning the last plant could still produce occasional, if genetically restricted, offspring. Past and ongoing climate change and habitat destruction is creating an extinction debt that will take a century or longer to play out. This raises the thorny question of how we identify the 'living dead' in the extinction waiting room.

Apocalyptic forebodings of uncertain future extinctions resonate because in spite of the acknowledged uncertainty in the numbers and time lags, plant extinctions are already upon us. Although hard to pin down, rates of plant extinction during the Anthropocene are about 10- to 100-fold higher than the natural background rate.[44] How does this measure up against past mass extinction events, and what can we learn about the prospects for life's survival and recovery? Palaeontologists are always keen to point out that 99% of all species that have evolved on Earth over the past 3.5 billion years have gone extinct.[45] Viewed over the immensity of geological time, extinction is the norm, with species having an average lifetime of 5–10 million years, depending on the group of organism; mammal species are less durable than plants, for instance. Yet sitting far above the background level of extinction are five well-documented so-called mass extinction events that occurred during the past 540 million years since complex plants and animals populated the planet. Sharp spikes in extinction rates far exceeding background levels characterize the so-called 'Big Five' mass extinctions, each representing the loss of 75% or more species. These past catastrophes in the history of life on Earth offer three important cautionary lessons for our current crisis.[46]

*Lesson 1.* Draw up a list of causes for each of the Big Five mass extinctions and it soon becomes apparent that a synergy of environmental factors pushed life over the edge, rather than any single event. The end-Permian mass extinction, 250 million years ago, wiped out nearly all life on Earth and coincided with vast episodes of volcanic activity in Siberia that injected huge amounts of carbon dioxide into the atmosphere, triggering global warming, and noxious ozone-destroying sulphurous gases.[47] The end-Triassic mass extinction, 201 million years ago, coincided with the birth of the Atlantic Ocean, as North America and Africa slowly drifted apart. The tectonic activity resulted in volcanism belching out huge amounts of carbon dioxide into the atmosphere. As the atmospheric carbon dioxide concentration soared, the hot, parched planet that followed wiped out biodiversity on land and acidified the oceans, to extinguish life in the seas.[48] Today, current rates of increase in the atmospheric carbon dioxide concentration, climate change, and ocean acidification far outstrip those experienced during any past extinction event. Humans have added to the mix with deforestation, habitat loss and fragmentation, expansion of agriculture, over-fishing, over-hunting, and pollution. Lesson one from Earth history is this: human activities are creating the perfect storm for the sixth mass extinction, priming the engine of species extermination.

*Lesson 2.* Rigorous meaningful comparisons between the magnitude of extinction events from hundreds of millions of years ago and the present are difficult. The fossil record is patchy, with much of the evidence for depletions in biodiversity obliterated, making rates of extinction hard to accurately pin down. Nevertheless, careful estimates suggest that in terms of magnitude, the Big Five trump current losses, with each losing between 75% and 90% of species.[49] In the 1990s, the World Conservation Monitoring Centre listed 592 plant species as having gone extinct since 1600. In 2016, The International Union for the Conservation of Nature Red List of Threatened Species listed fewer than 150 extinct species of plants. Granted the strong suspicion of undocumented extinctions, these ballpark estimates of magnitude for plants do not yet qualify as mass extinctions in the palaeontological sense. It is on the question of rates that things become concerning, with the best estimates of plant species extinction for the Anthropocene far exceeding those calculated for the Big Five, if we scale past mass extinction rates to the same window of time.[50] Lesson two from Earth history is that unless we relieve the pressures that are pushing today's species to extinction, we will propel the world towards the sixth mass extinction within a few generations. For vertebrates—mammals, birds, fish, amphibians, and others—it may already be underway, with recent declines in their diversity unprecedented in the past 65 million years.[51] Plants have so far fared better, but may be sitting in the extinction waiting room for longer than animals, for reasons we have already discussed.

*Lesson 3.* Extinctions from the past provide a rough guide to the recovery time of the biosphere, following catastrophic losses. On average, it takes anything from hundreds of thousands of years to tens of millions of years for ecosystems to recover, reorganize, and bounce back to pre-extinction levels of biodiversity. After the end-Permian mass extinction, tropical forest and coral reef diversity remained depleted for ten million years or so, and global biodiversity took a hundred million years to recover.[52] After the end-Cretaceous mass extinction event, 65 million years ago, that famously saw off the dinosaurs, the diversity of life in the seas took millions of years to rebound.[53] Lesson three from Earth history is that if our actions drive life towards a sixth mass extinction, we are likely sealing the fate of biological diversity for millions of years.[54]

By 1992, nations of the world had recognized the urgency of the situation and signed up to the UN Convention on Biological Diversity. By 2002, 193 parties had pledged to lower rates of biodiversity loss substantially by 2010. Endorsed by the World Summit on Sustainable Development, this goal became incorporated into the UN Sustainable Development Goals for 2030. But 2010 came and went, and despite the political posturing over bold commitments to biodiversity conservation, the world saw no significant slowing of biodiversity decline between 1970 and 2010.[55] Prospects for the diversity of plant and animal species continue to remain grim. A chink of light appeared in October 2010 when the Aichi Targets of the Strategic Plan for Biodiversity 2011–2020 became enshrined in the UN's Convention on Biological Diversity to halt biodiversity loss.[56] One of these targets sought formal protection for 17% of the terrestrial world. In the same year, the UN updated the Global Strategy for Plant Conservation with the aim of protecting 60% of plant species. Are these two aspirational goals compatible? The good news is that we are currently protecting ~13% of the global land area. The bad news is that the protected areas are not in the right places to protect 60% of plant species, with a bias towards land in cool, high, or dry places or places far away from people and with little economic value.[57] Conservationists urge preservation of tropical and sub-tropical islands, moist tropical and sub-tropical forests, and Mediterranean ecosystems that hold high numbers of endemic plant species. Other approaches to conserving biodiversity are necessary and under development.

The continued lack of co-ordinated action to halt biodiversity decline, despite the repeated setting of these international targets,[58] is a great tragedy of our time.

The voice of conservationists has no particular standing with politicians, and few seem to take notice of species disappearing each year, or the rapidly climbing rate of their disappearance. Curves depicting rising global temperatures and atmospheric carbon dioxide concentrations over the past fifty years seem to have clarity in demonstrating the reality of climate change; tabulated lists of threatened species appear somehow to be less compelling.[59] We are continually losing opportunities for the future, and committing biodiversity to extinction during the course of this century. As the Harvard evolutionary biologist Edward Wilson points out, this is a sin, for which future generations are least likely to forgive us, because there is no way to get them back once they are consigned to the dustbin of history. The pioneering genome scientist J. Craig Venter and his team may be on the verge of synthesizing new life forms by assembling and inserting a synthetic genome into a bacterial cell,[60] but, ironically, we are failing to preserve the biodiversity we already have.

Yet saving the world's floras and fauns from the threat of climate change is far from straightforward and these simple statistics hide the complex challenges facing conservationists. A good example is the Californian Floristic Province (CFP). The CFP is a stunning world-renowned biodiversity hotspot with a distinctive flora composed of over 5000 native plant species, with more than 2000 endemics. Species–area relationship analyses, together with the projected changes in climate over the next century, predict that some 60% of species could experience drastic contractions in their ranges. Yet California's varied terrain, with mountains and desert areas, means that species may be forced to move in very different directions, causing a break-up of the composition of present-day floras. Quite what might happen will be sensitive to actual rates of plant migration, and raises the question of whether we should intervene to help plant dispersal as part of a conservation strategy. If not, weedy species with short lifetimes and easily spreading seeds are poised ready to fill the gaps in California, and those created elsewhere in the world. This presents a dilemma to conservationists: should they prioritize areas, refugia, where we expect plant diversity to peak in the face of future climate change? These 'future' refugia may contain a disproportionately high number of threatened species that could persist, offering valuable opportunities for conservation. Or should conservationists devote time and effort to protecting mountainous areas harbouring species threatened by drastically shrinking ranges? In the end, the number of species projected to survive climate change may depend on their ability to disperse into suitable habitats and climates. In a world with increasingly fragmented habitats, landscape connectivity for allowing species to migrate is crucial.

Another conservation perspective that shifts emphasis away from the overriding preoccupation with preserving rare species in threatened diversity hotspots comes from thinking about evolutionary history. The great Russian-born evolutionary biologist Theodosius Dobzhansky (1900–1975) may have been indulging in some shameless self-promotion when he famously proclaimed that 'nothing in biology makes sense except in the light of evolution', but few scientists would disagree with the sentiment. Now conservationists are also embracing the need to recognize the importance of the evolutionary history of species, the accumulated baggage of their past. History matters in the delicate game of surviving threats like climate change and habitat fragmentation because where you sit on the green tree of plant life can determine your risk of extinction. Your position on the tree also says something about the amount of evolutionary history your extermination will wipe out. Consider that, if a species is at risk of extinction, then its close relatives with similar ecological and physiological characteristics will also have a higher than average chance of being at risk too. The extinction of species clustering together on the green tree of life could jeopardize the existence of entire families and genera of plants.[61]

Saving evolutionary history itself might be a valuable focus for the conservation movement. We could target areas preserving patterns of biodiversity and the evolutionary processes that generated the patterns, bringing pre-emptive conservation strategies within reach. Combine the evolutionary histories of species with their geographical distributions and you can discover the 'cradles' of diversity (regions where diversity is generated), and can distinguish them from 'museums' of diversity (regions where it persists).[62] Take the floras of the Cape Province of South Africa, for example. It is an area of less than 90 000 km$^2$ and, like the Californian Floristic Province, represents another extraordinary biodiversity hotspot, containing more than 9000 plant species, 70% of which are endemic.[63] Most of the threatened plant species belong to relatively young, fast-evolving lineages making up the tips of the green tree of life.[64] Put another way, rapidly diversifying groups of plants contain the more vulnerable species. Because much of the present-day plant diversity in the Western Cape results from recent speciation events, plant extinctions will erase relatively little of the evolutionary history of this biodiversity hotspot. This is not to reject areas in the region with recently diversifying floras as conservation targets. Rather, it points to the need for including other regions in conservation schemes to maximize the potential for preserving evolutionary history.[65]

As conservationists face these sorts of complex issues, the sceptic, and indeed the heretic, might ask—does it matter? To which we surely offer the indignant reply that we have a profound moral imperative and ethical obligation to preserve Earth's rich biological diversity for future generations. After all, the estimated 7–12 billion species of organisms (other than bacteria or viruses) with which we share the planet are astonishing products of the same evolutionary process that gave rise to us. There is also another, pragmatic response to this question to consider, that of preserving essential 'ecosystem services' provided by nature. Here, in this utilitarian view of nature, plants often take centre stage. They underpin high-profile ecosystem functions, including the things we take for granted such as food, carbon sequestration, water, timber, fuel, soils, nutrient cycling, pollination, and pharmaceuticals.

Over ten years ago, Robert Constanza and colleagues controversially attempted to estimate the value of ecosystem services worldwide.[66] Acknowledging substantial uncertainties in such an exercise, Constanza put a crude initial price on the 'services' provided by the biosphere as a whole of around $33 trillion per year. As he points out, 'The economies of the Earth would grind to a halt without the services of ecological life support systems, so in one sense their total value to the economy is infinite'. Putting a price on nature is an unnervingly anthropocentric view of the world, mistakenly implying nature is here to serve us. Are we comfortable with reversing over 500 years of scientific thinking from the greats like Copernicus, Galileo, Hutton, and Darwin who each played a role in the gradual disassembly of the seemingly special and central position of Earth-bound humans in the Universe? Probably not, but the pragmatic view is that without the bold step of valuing ecosystem services, we cannot begin to engage policymakers, engineers, and other scientists whose decisions threaten the sustainability of the planet. According to economists, the price to pay for the erosion of nature's ecosystem services could be in the region of $2–5 trillion per year.[67]

Not surprisingly, biodiversity underpins the ecosystem services on which we depend for our very existence. If it falls below some safe threshold or limit, then the wide range of services which are central to our health and well-being, such as crop pollination, and regulation of the global carbon and water cycles, are critically threatened. Framed in this context, the most comprehensive analysis of biodiversity change to date suggests that 58% of the world's land surface has fallen below the safe threshold.[68] Hold on a minute, you might cry. If such a large proportion of land has already passed below the safe threshold for biodiversity loss, why have we not already noticed negative effects on human society and wellbeing? The Stanford ecologists Paul and Anne Ehrlich offer an engineering metaphor by way of explanation in their book *Extinction: The Causes and Consequences of the Disappearance of Species.* Published in 1981, decades before the conceptualization and economic valuing of ecosystem services, they neatly illustrate how uncertainty in biodiversity loss relates to the functioning of ecosystems by likening it to the loss of rivets from an aeroplane wing. How many rivets are we comfortable with popping out before refusing to fly? 'Ecosystems, like well-made airplanes, tend to have redundant subsystems and other features that permit them to continue functioning after absorbing a certain amount of abuse. A dozen rivets, or a dozen species, might never be missed. On the other hand, a thirteenth rivet popped from a wing flap, or the extinction of a key species involved in the cycling of nitrogen, could lead to a serious accident.'[69] In this context, we should recall American conservationist Aldo Leopold's (1887–1948) great quote, 'the first rule of intelligent tinkering is to save all the cogs and wheels'.

Another way to think of how our destruction of nature degrades ecosystem services is the 'ecological footprint' concept. Conceived by Mathis Wackernagel and William Rees, an ecological footprint provides a metric for calculating the human pressure on planet Earth.[70] The idea is that natural ecosystems generate a finite amount of raw materials, like wood from trees, crops, fish, and so on, and, in return, absorb a finite amount of waste, like carbon dioxide. The ecological footprint measures humanity's demand on nature as the balance between these two. It asks the question: how fast are we consuming resources compared to the capacity of nature to absorb our waste and regenerate them? An ecological footprint is the area of the Earth's surface needed to produce all the resources that a population of humans consumes, and to absorb the waste they generate. The concept is hugely influential, with governments, businesses, and institutes worldwide adopting it to monitor resource use and advance sustainable development.

As an environmental accounting tool, ecological footprint calculations have much to recommend them. They are versatile and scalable for individuals, businesses, cities, nations, barrels of oil, or even the globe. You can estimate how much land it takes to support your own personal lifestyle by visiting the on-line ecological footprint calculator and answering a series of questions about how you live your life.[71] Take the challenge and you may discover it can be a sobering experience. If everyone had the same sybaritic lifestyle as affluent western Europeans, we would require the regenerative capacity of 2.6 planet Earths per year to support ourselves.[72] The calculator also forcefully brings home to us the point that our personal decisions matter to the environment. Individuals can make a difference. How much do you drive your car or opt to use public transport? How fuel-efficient is your car? Do you diligently recycle products? Have you insulated your home effectively to conserve energy? These types of questions highlight opportunities for re-orientating our lifestyles and reducing our negative impacts on the environment.

Across the world, not enough of us are making, or having the opportunity to make, the right choices. In consequence, the ecological footprint for the human population, all 7.6 billion of us, is substantial. Advances in science, technology, and medicine, and our merciless overexploitation of Earth's natural resources, ensure humanity currently uses the equivalent of 1.5 planet Earths to provide the resources consumed and to absorb our waste. It now takes the Earth one and half years to regenerate what we use in a year. Human demands on nature exceed the capacity of photosynthesizing ecosystems on land and in the sea to supply them or regenerate. Without immediate action, this overshoot will get worse. Moderate United Nations scenarios suggest that, if current population and consumption trends continue along the 'business as usual' trajectory, by the 2030s we will need the equivalent of two Earths to support us. That is a major problem. It means humanity is systematically liquidating the Earth's assets, turning valuable natural resources into waste much faster than nature can regenerate them. The only way to avoid such a dramatic overshoot in demand relative to supply is to take action by shrinking the aggregate demands of humans on the environment, and moving towards a sustainable future.

Stark visions of a grim future haunted by depauperate biodiversity and diminished ecosystem services raise the pressing question of how we address the root causes of the problem: rapid human population growth, secure food production, per capita consumption and production, and climate change. Developing sustainable food and energy sources that safeguard biodiversity and climate by shrinking humanity's ecological footprint are key challenges. Jon Foley, current Director of the California Academy of Sciences, spends more time than most thinking about how we might resolve the dilemma of enhancing sustainable food production, without further collateral degradation of the environment.[73] He argues that we have to transform global agriculture fundamentally to deliver food security. At the same time, we must improve distribution and access to food given the tragic irony that we could already feed everyone on the planet even without increasing crop productivity. Approximately a billion people go to bed at night hungry, chronically malnourished, not because we cannot produce enough food, but because of failures by society to solve its effective distribution.[74] In a just and fair world where there is enough food to go around, no one should be undernourished. Most of these hungry people live in Asia and sub-Saharan Africa, the only regions where over 30% of people are malnourished, and the numbers are increasing. Non-governmental organizations such as Oxfam continue to lobby governments, expose bad business practices, and generate media interest about this critical issue, whilst supporting grassroots movements for change. The bottom line is that the global food system needs fixing with investment, political will, and a coalition of stakeholders to promote trade.[75]

Ways forward to protect biodiversity are easy to identify in principle, but difficult to achieve in practice. Halting our relentless expansion of agricultural land into biodiverse tropical forests is an obvious place to start. Many regions cleared for agriculture in the tropics have crop yields well below those planted in temperate regions anyway.[76] High-yielding tropical crops like sugarcane, oil palm, and soybean do not contribute much to the world's total food production. Given UN forecasts of world population growth, agricultural yields will need to increase with a changing global climate. Could this be achieved with existing croplands without deforestation? Existing croplands are not being used efficiently; many regions are underperforming landscapes with below-average crop yields.

Opportunities exist in Africa, Latin America, and Eastern Europe for boosting productivity by improving irrigation and road networks that transport labour and inputs to farmland, as well as the cultivation of improved crop varieties. According to Foley, if we could raise the yields of just 16 key food and animal feed crops in these regions to within 75% of their full potential, global food production would increase by 58%. If we raise crop yields to within 95% of their full potential, food production would increase by an additional 2.3 billion tonnes. Radical adaptation of agricultural practices is part of this solution. Precision irrigation and organic farming systems that recycle crucial nutrients have to be part of the mix (Chapter Seven). Organic farming, once thought to be the wrong call,[77] may yet play its part in an integrated solution. Critics have argued that lower yields means organic farming requires more land to produce the same amount of food as conventional agriculture and that this undoes environmental benefits.[78] But yields of well-managed organically grown crops average 75–80% of the same crops under conventional management.[79] Opportunities exist for improving water and nutrient efficiency of agriculture, without reducing food production, by reducing excessive fertilizer use, improving manure management, and capturing excess nutrients through recycling. Worldwide, 10% of the world's crop lands account for 32% of the excess nitrogen surplus and 40% of the phosphorus surplus. Resolving what is called the Goldilocks nutrient problem—many regions of the world have too much or too little fertilizer, few are just right—would improve the efficiency of nutrient use and reduce agriculture's damaging environmental impacts.[80]

Improving yields will require genetic improvements in crops. This is nothing new. We have been breeding for increased yields and disease resistance in crops for centuries, by crossing crop varieties with each other and related species. There is no panacea for solving the global challenge of sustainable and secure food production in a changing climate and rising population, but genetic technologies can help meet future needs.[81] They offer ways of producing higher-yielding crops, crops with added nutrients (including 'golden rice' with added beta carotene, from which our bodies make vitamin A), and crops that are resistant to drought, pests, and herbicides. Objections raised by campaigners to transfer of non-plant genes to crops to provide herbicide resistance lack solid scientific evidence, and there are few situations where genetically modified crops pose a threat to biodiversity or the environment.[82] In fact, the opposite is true. Traditional farming systems with high applications of chemicals to crops often pose a greater threat.

Revolutionary changes in our diet will also be necessary, no matter how unpalatable. If we all ate less meat, it would dramatically reduce the agricultural land area put to the plough, and improve our health by lowering our risk of obesity, coronary heart disease, and intestinal cancers.[83] Shifting from grain-fed beef to pasture-fed beef or poultry could enhance crop capture of sunlight by freeing up land. Chickens convert grain into protein three times more efficiently than cattle. Weaning ourselves off meat altogether could release over 2700 million hectares of pasture, and a hundred million hectares of cropland.[84] Although unlikely to be popular, that scenario is a win–win situation. Abandoned land could become revegetated, ideally with diverse forests that sequester carbon, and the abandonment of livestock farming would reduce the release of potent greenhouse gases—methane and nitrous oxide—into the atmosphere.

Transforming agricultural practices and dietary habits are necessary steps for alleviating pressures on global ecosystems of plants and animals, but are unlikely to be sufficient. We urgently need to address the climate change issue. Nations of the world agreed and recognized the need to limit fossil fuel emissions to avoid dangerous climate change by signing up to the UN 1992 Framework Convention on Climate Change. Curbing greenhouse gas emissions will not only help avoid seeding irreversible climate change but will also help avoid seeding irreversible loss of species diversity. James Hansen, former Director of NASA's Goddard Institute of Space Studies, New York, and a tireless advocate of the need for action on climate change, has shown that the UN's poorly defined objectives are already out of date. Atmospheric concentration of greenhouse gases and global temperature rise since the 1970s have overshot safe limits.[85] Earth is now warmer than the last interglacial interval in Earth's history. Called the Eemian period, this interglacial occurred 130 000 to 115 000 years ago, when sea level was 6–9 metres (20–30 feet) higher than today. Accelerated melting of the major ice sheets on Antarctica and Greenland is already underway and substantial, multi-metre sea-level rise might be on the cards in the coming century.[86] The consequences of future sea-level rise worsen as population sizes in coastal regions increase. Bangladesh, for example, is smaller than Washington State but has a population half that of the USA. Its population doubled from 80 million in 1975 to 160 million in 2013, and is forecast to increase further to 200 million by 2050. Ice-sheet melt and sea-level rise are amongst the most threatening features of climate change. Even if we follow a medium-level mitigation scenario, which right now feels optimistic, the long-term consequences for sea-level rise could be disastrous. Mass migration of hundreds of millions of people may result from the flooding of coastal megacities (those with a population of 10 million or greater), and submergence of heavily populated low-lying regions, creating an unprecedented humanitarian crisis within the next century or two.[87]

The bottom line is that meeting UN climate targets and protecting nature means leaving fossil fuels in the ground. We have extracted and burned only a small fraction of the total available fossil fuel resources. Since the onset of the Industrial Revolution, this amounts to 374 billion metric tonnes of carbon from fossil fuels (coal, oil, and gas) extracted from easily accessible reserves. Continued exploitation of less-accessible unconventional fossil fuels will prove more costly to planetary ecology, with a larger ecological footprint. Extraction will require more inputs, more land, more water, and more energy, and produce more waste. Coal mining is shifting from predominantly subsurface to surface mining, with forest cover and topsoil destroyed and excess rock removed with explosives. Oil mining is exploiting fields of decreasing size that are more widely geographically dispersed, leading to greater fragmentation of ecosystems by road and rail networks and pipelines. In Canada, the area of land required per barrel of oil produced increased 12-fold between 1955 and 2005.[88] Gas mining is being pursued by unconventional means through fracking, with only a rudimentary understanding of the environmental and ecological consequences.

The urgency of the situation is real and understood.[89] Carbon dioxide emissions from burning fossil fuels are locking in our inevitable rendezvous with a warmer planet, diminished in diversity. Left unchecked, global temperatures could eventually increase by 3–4°C or more by 2100, increasing extinction risks for plants and animals, which climb with every degree of warming. Yet despite widespread recognition of the risks, high fossil fuel emissions continue; emissions in 2014–2016 were the highest in human history.[90] The major problem is that governments subsidize heavily the costs of fossil fuels, making them the cheapest reliable source of energy. So-called 'cap and trade' schemes fail because they simply allow developed nations to pay developing nations to continue their emissions. Hansen and others[91] argue that the most effective way to slow emissions is a steadily rising carbon fee that makes fossil fuels pay their true cost to society with proceeds from this carbon tax returned to citizens. Dividends generated by the tax would reward people for reducing their collective carbon footprint. Economists suggest the carbon dividend would grow over time as the carbon tax rate increases, creating a positive feedback loop: greater climate protection → greater individual dividend payments → less fossil fuel use → greater climate protection. Implemented by a few major powers, a transparent rising tax on fossil fuels represents a policy instrument that could refocus efforts to improve energy conservation, and drive government support for the development of clean energy technologies. The idea is gaining support, with citizens' groups advocating implementation to create the political will to build climate solutions.[92]

Unfortunately, political leaders and policymakers are not listening. If governments fail to show strong leadership in transforming food and energy systems, climate change and locked-in extinction debts guarantee future generations will inherit a planet in peril. This raises profound moral questions about intergenerational justice and the fundamental rights of young people to inherit a habitable planet whose biodiversity is not threatened with mass extinction and whose climate is not unstable. The possibility is not remote. It is already on our doorstep. Given what is at stake, society's attitude to nature has to change, with a greater emphasis on the urgent need for action. The headline message—sustainable food and energy production, without collateral damage to the planet—has to resonate with young minds. We have to inspire young people to get involved with decisive actions to change the future course of the Anthropocene.

Raven (**Figure** 27) provides a case study for how engaging young people can successfully play out in the long run. Not long after moving to San Francisco from Shanghai in 1937, he became hooked on the natural world. 'I always say that all children are interested in nature through observing nature and collecting, and we mustn't knock it out of them.' By the age of five, he had joined a local natural

**Figure** 27 Peter H. Raven, former Director of the Missouri Botanic Garden and pioneering plant conservationist.

history club, read about butterflies and beetles, and made the great discovery that male and female house sparrows were the same species. Between 1950 and 1956, aged 14 to 20, he spent six weeks every summer camping in the Sierra Nevada Mountains of western North America working with naturalists and developing his knowledge of the natural world. After moving through the academic system he arrived at Stanford University, before becoming the Director of the Missouri Botanic Garden in 1971. Thirty years later, President Clinton awarded him the National Medal of Science for his contributions to the fields of biodiversity and the environment.

How do we get their attention? Raven suggests, 'We could emphasize that two thirds of the people of the world depend directly on plants for their medicines; and for the rest of us, plants supply about a quarter of our drugs, either directly or through manufacturing the compounds they produce naturally, [and] another quarter come from microbes and fungi. Or we could point out that biodiversity collectively makes possible our continued existence on Earth; it makes our lives possible. Our soils, the water we drink, the nature and quality of the air we breathe, the beauty of our lives, and importantly our prospects for the future, all depend on maintaining the rich biodiversity that we have inherited, and for which are clearly responsible.' As he points out, 'We are but one species that grew up here, what right do we have to destroy everything on Earth? If you were a visitor of Earth from another planet, the thing that would strike you the most is the wonder and beauty of the fantastic organisms on Earth. You wouldn't start destroying them; you'd want to spend your whole life learning about them.' At least that would be the hope for most of us, but not everybody is likely to be convinced. On learning that an endangered breed of whales was being systematically turned into soap, the writer Kingsley Amis remarked that 'it sounds like quite a good way of using up whales'. Obviously, we can make arguments for saving species diversity but 'the fact of the matter is they don't convince people. They don't carry the weight of argument and I don't know exactly why not because they sound as if they would. Perhaps they're too intellectual?'

Raven's uncompromising diagnosis of the prospects for life on Earth hinges on a clear appreciation of the stark realities of the situation.[93] 'Politically what can you say? There is nothing a politician could legitimately run on to get elected. Every politician is elected by saying you are all going to be wealthy, that is the dream. They forget that England and America, and all these other countries, are living in a complete bubble way above the capacity of the Earth to support them. We can't go on consuming more than the total productivity of the Earth. The solution has a serious, moral underpinning. In a sense the golden rule, that really does say it all, is that if we could really love one another worldwide, I'm sure that we could solve these problems. But we are so far from that. We are wary and looking out for ourselves and wanting to accumulate things and enrich our lives—selfishness was highly adaptive pre-agriculture, but the more people that adopt it the less adaptive it is. Selfishness is like a cancer. Somebody once said to Gandhi, "What do you think of Western civilization?" He said it "sounds like a good idea, someone ought to try it". I don't think that ought to be given short shrift because in a way that actually underlies everything. We need to work collectively.'[94]

The situation recalls J.G. Ballard's prescient missives from suburbia that have occasionally punctuated our thinking throughout this book. Ballard repeatedly warned of the social fallout of post-war urbanization where people become obsessed with themselves and their security. Ballard, like Raven, was born in Shanghai, China, but stayed far longer. In 1943, his family was interned in the Lunghua Civilian Assembly Camp until the Hiroshima and Nagasaki atom bombs ended the war. Environmental change features heavily in his early apocalyptic novels (exemplified by *The Drowned World* and *The Drought*) and warning us of the need to build a better society was central to his agenda. 'As living standards continue to rise, as they have done since the war—and I'm sure living standards will, on the whole continue to rise—people have got more to lose. You know, they've packed their homes with high-tech electronic gear... not to mention things like jewellery. So one gets a strangely interiorized style of living where you switch off the outside world. You do this by treble-locking your front door and switching on the alarm system, and then you retreat and watch videos of the World Cup. And that's not a good recipe for a healthy society.'[95] Or a healthy planet, we should add.

Humanity stands at a crossroads. Preserving the lush green planet we enjoy requires protecting biodiversity and planetary climate and this ought to be a top priority for citizens, nations, and governments of the world. We are completely dependent on plants now and will be in the future. Only with co-operation between nations, and by taking responsibility for our own actions, can we develop global solutions for a sustainable, more equitable world. The window of opportunity is closing as the era of consequences dawns. If we make the right choices in the coming decades, there is a chance of avoiding the sixth great wave of extinction in the history of life on Earth, one that is of our own making. If we fail to address the planetary crisis, a far worse alternative lies ahead.[96] Fortunately, science tells us there is still time to take action and reject that grim alternative future of sending our emerald planet back to the drab world of the distant past.

# SIMPLIFIED GEOLOGIC TIMESCALE FROM THE CAMBRIAN

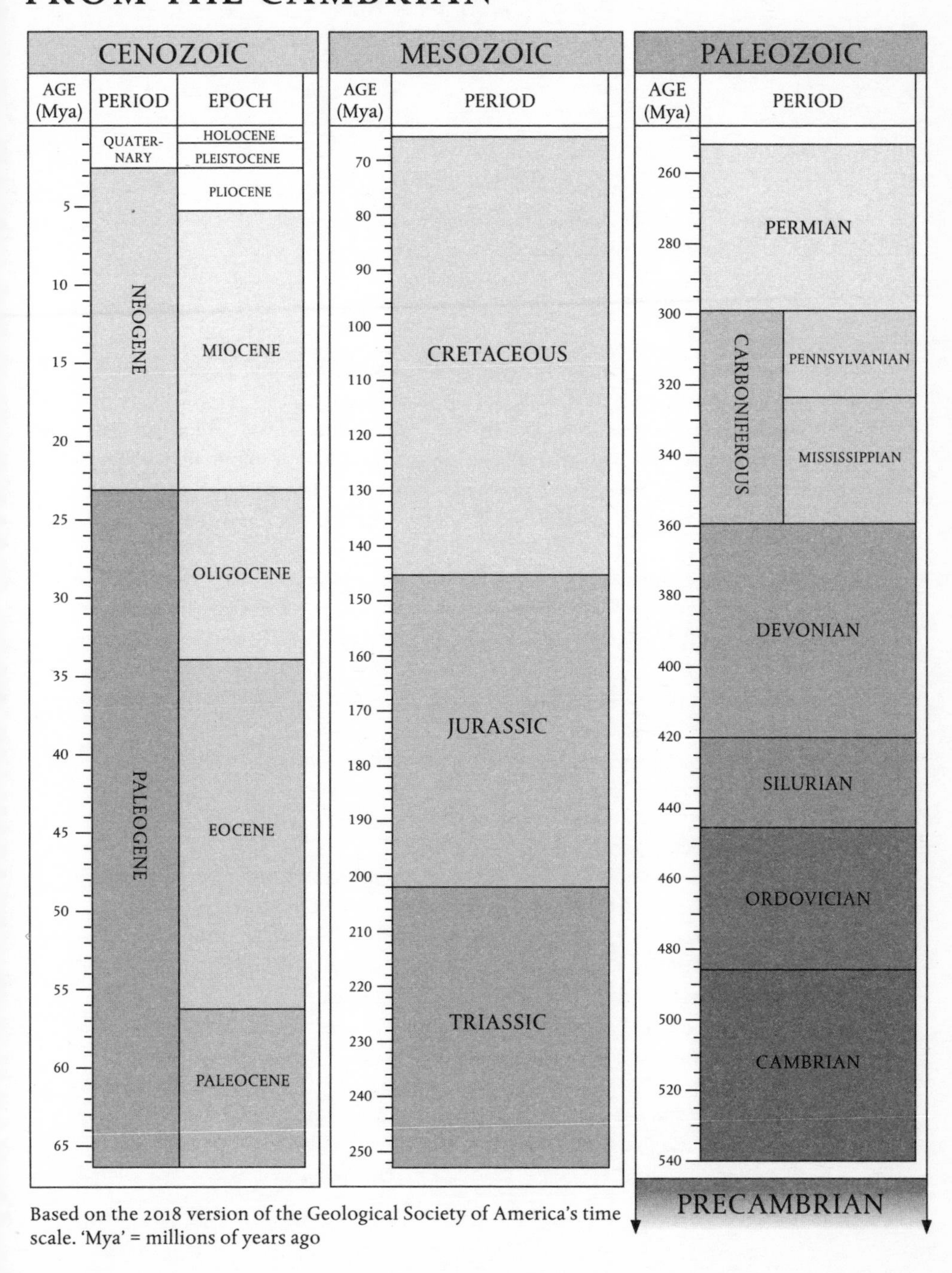


Based on the 2018 version of the Geological Society of America's time scale. 'Mya' = millions of years ago

# ENDNOTES

## 1. All flesh is grass


1. Christopher, J. (1956) *The Death of Grass*. Penguin, London.
2. Ballard, J.G. (1965) *The Drought*. Fourth Estate, London.
3. McCarthy, C. (2006) *The Road*. Picador, London.
4. Ceres was discovered by the Italian astronomer Giuseppe Piazzi (1746–1826) on 1 January 1801.
5. See Sleep, N.H., Bird, D.K. & Pope, E.C. (2011) Serpentine and the dawn of life. *Philosophical Transactions of the Royal Society*, **B366**, 2857–69.
6. Although we know little about the earliest form of life, we are no longer tied to the intriguing yet ultimately untenable suggestion of the evolutionary biologist J.B.S. Haldane (1892–1964) that a primordial soup may have provided the basis for the origin of life, as given in Haldane, J.B.S. (1929) The origin of life. *Rationalist Annual*, **3**, 3–10.
7. See Martin, W. & Russell, M.J. (2007) On the origin of biochemistry at an alkaline hydrothermal vent. *Philosophical Transactions of the Royal Society*, **B362**, 1887–1925. Martin, W. et al. (2008) Hydrothermal vents and the origin of life. *Nature Reviews in Microbiology*, **6**, 805–14.
8. Früh-Green, G. L. et al. (2003) 30,000 years of hydrothermal vent activity at the Lost City vent field. *Science*, **301**, 495–8.
9. Lane, N. et al. (2010) How did LUCA make a living? Chemiosmosis in the origin of life. *BioEssays*, **32**, 271–80.
10. Field, C.B. et al. (1997) Primary production of the biosphere: integrating terrestrial and oceanic components. *Science*, **281**, 237–40.
11. Vermeij, G.J. & Grosberg, R.K. (2010) The great divergence: when did diversity on land exceed that in the sea? *Integrative and Comparative Biology*, **50**, 675–82. May, R.M. (1994) Biological diversity: differences between land and sea. *Philosophical Transactions of the Royal Society*, **B343**, 105–11. Grosberg, R.K., Vermeij, G.J. & Wainwright, P.C. (2004) Biodiversity in water and on land. *Current Biology*, **22**, R900–3.
12. Labandeira, C.C. (2005) Invasion of the continents: cyanobacterial crusts to tree-inhabiting arthropods. *Trends in Ecology and Evolution*, **20**, 253–62.
13. Clack, J.A. (2007) Devonian climate change, breathing, and the origin of the tetrapod stem group. *Integrative and Comparative Biology*, **47**, 510–23. For a popular treatment dealing with all aspects of vertebrates coming ashore, see Clack, J.A. (2012) *Gaining Ground: The Origin and Evolution of Tetrapods (Life of the Past)* (2nd edn). Indiana University Press, USA.

14. Seward, A. C. (1914) Antarctic Fossil Plants. British Antarctic ('Terra Nova') Expedition, 1910. British Museum Natural History Report. *Geology*, **1**, 1–49.
15. Wandersee, J.H. & Schussler, E.E. (1999) Preventing plant blindness. *American Biology Teacher*, **61**, 82–6. Wandersee, J.H. & Schussler, E.E. (2001) Toward a theory of plant blindness. *Plant Science Bulletin*, **47**, 2–9.
16. In an effort to combat humanity's appalling plant blindness, and win the battle for young minds, Wandersee and Schussler produced a poster, endorsed by the Botanical Society of America and distributed to more than 20 000 schools, with the hope of recruiting the next generation into the exciting science of botany. They maintain an exciting website for their visual cognition research development laboratory (http:// www.15degreelab.com). The laboratory takes its name from the observation that most people typically view objects that lie between 0° and 15° below eye level. We really do need to look up more to overcome plant blindness.
17. In a fascinating North American study, university botany students were shown 14 pairs of animal and plant images, and their recall of each image was determined after undertaking a distracting task. The results showed that they consistently recalled the images of animals more accurately, and more often, than those of plants. See Schussler, E.E. & Olzak, L.A. (2008) It's not easy being green: student recall of plant and animal images. *Journal of Biological Education*, **42**, 112–18.
18. New, J., Comsides, L. & Tooby, J. (2007) Category-specific attention for animals reflects ancestral priorities, not expertise. *Proceedings of the National Academy of Sciences, USA*, **104**, 16598–603. See also the commentary: Ohman, A. (2007) Has evolution primed humans to 'beware the beast'? *Proceedings of the National Academy of Sciences, USA*, **104**, 16396–7.
19. Gee, H., Howlett, R. & Campbell, P. (2009) 15 Evolutionary Gems. *Nature*, **457**, doi: 10.1038/nature07740. Available at http:// www.nature.com/evolutiongems.
20. Bower was following original research by the Czech botanist Ladislav Josef Celakovsky (1834–1902). Sadly, we know little about Celakovsky's life.
21. Walton, J. (1948) Obituaries. Prof F.O. Bower, F.R.S. *Nature*, **161**, 753–4.
22. Gribbin, J. (2003) *Science: A History 1543–2001*. Penguin Books, London.
23. YouTube is currently offering a memorable reading by David Horovitch of this haunting tale: http://www.youtube.com/watch?v=gjeEaviwUYQ
24. Frank's botanical and mycological colleague Anton de Bary (1831–1888) devoted much of his life to studying fungi and greatly expanded the concept; see de Bary, H. (1879) *Die Erscheinung der Symbiose*. Strasberg, Germany. As one of a new breed of active continental botanists using advanced experimental techniques to understand how nature works, de Bary attracted many scientific visitors eager to learn his approach. Bower spent a full academic year with him in Strasburg in 1879, where he mastered German and attended the great professor's lectures to improve his mind, and general education in botany; see Lang, W.H. (1949) Fredrick Orpen Bower. *Obituary Notices of the Royal Society*, **6**, 347–74.
25. Kidston, R. & Lang, W.H. (1921) On Old Red Sandstone plants showing structure, from the Rhynie chert bed, Aberdeenshire. Part V. The thallophyta occurring in the peat-bed; the succession of the plants through a vertical section of the bed, and the conditions of accumulation and preservation of the deposits. *Transactions of the Royal Society of Edinburgh*, **52**, 855–902.

26. Bierbaum, R.M. & Raven, P.H. (2007) A two-pronged climate strategy. *Science*, **316**, 17.
27. Dawkins, R. (2009) *The Greatest Show on Earth: The Evidence for Evolution*. Bantam Press, London.

## 2. Fifty shades of green

1. Belnap, J. (2003) The world at your feet: desert biological soil crusts. *Frontiers in Ecology and the Environment*, **1**, 181–9.
2. Death Valley in the Mojave Desert is the hottest place in North America—summer temperatures regularly soar above 120°F (49°C).
3. Wellman, C.H. & Strother, P.K. (2015) The terrestrial biota prior to the origin of land plants (Embryophytes): a review of the evidence. *Palaentology*, **58**, 6501–627.
4. Cardon, Z.G., Gray, D.W. & Lewis, L.A. (2008) The green algal underground: evolutionary secrets of desert cells. *Bioscience*, **58**, 114–22.
5. An excellent account of Linnaeus is given in Gribbin, J. (2003) *Science: A History 1543–2001*. Penguin, London.
6. Before the molecular sequencing revolution, classification was undertaken by looking for features that are similar or different, the principle of inheritance by which, for example, we share similar characteristics to our parents or grandparents. The fact is that today almost all evolutionary trees are based entirely on 'features' observed by DNA sequencing, and for good reason. The morphological characters of organisms, like those studied by Linnaeus and Darwin, were often confused by confounding issues of convergent evolution. Convergent evolution is the process by which natural selection favours similar adaptations (or characters) in different species not because they are closely related to each other, but because they equip those species similarly for survival in particular habitats. Chapter and verse on convergent evolution can be found in Conway-Morris, S. (2003) *Life's Solutions*. Cambridge University Press, Cambridge.
7. Cheng, C., Bowman, J.L. & Meyerowitz, E.M. (2016) Field guide to model plant systems. *Cell*, **167**, 325–39.
8. See Butterfield, N.J. (2000) *Bangiomorpha pubescens* n. gen., n. sp.: implications for the evolution of sex, multicellularity, and the Mesoproterozoic Neoproterozoic radiation of eukaryotes. *Paleobiology*, **26**, 386–404).
9. Delwiche, C. & Cooper, E.D. (2015) The evolutionary origin of a terrestrial flora. *Current Biology*, **25**, R899–910.
10. Margulis, L. (1981) *Symbiosis in cell evolution*. F.H. Freeman, New York. Sadly, Margulis died in 2011. For a concise overview of her revolutionary contribution to cell biology and evolution, see Lake, J.A. (2011) Obituary. Lynn Margulis (1938–2011). *Nature*, **480**, 458.
11. Mereschkowsky, C. (1905) Über Natur und Ursprung der Chromatophoren im Pflanzenreiche. *Biol. Centrabl.*, **25**, 593–604.
12. Lake, J.A. (2011) Obituary. Lynn Margulis (1938–2011). *Nature*, **480**, 458.
13. The classic paper is: Schwartz, R.M. & Dayhoff, M.O. (1978) Origins of prokaryotes, eukaryotes, mitochondria, and chloroplasts. *Science*, **199**, 395–403.
14. For a review, see Keeling, P.J. (2010) The endosymbiotic origin, diversification and fate of plastids. *Philosophical Transactions of the Royal Society*, **B365**, 729–48.

15. If you think this seems far-fetched, consider the strange case of the herbivorous sacoglossan sea-slugs, minuscule molluscs that dwell on the Great Barrier Reef and elsewhere in the world's oceans. These solar-powered sea-slugs have a distinctive bright green appearance that comes from the chloroplasts of the green algae they eat but fail to fully digest, see Händeler, K. et al. (2009) Functional chloroplasts in metazoan cells—a unique evolutionary strategy in animal life. *Frontiers in Zoology*, **6**, 28. In other words, these little green molluscs are kleptoplasts, 'stealing' and retaining chloroplasts from their algal food to camouflage themselves and avoid predation. The chloroplasts also provide the slugs with additional energy in the form of sugars manufactured by photosynthesis, a function that seems to be especially important for enhancing their survival when times are hard and the algae they feed on scarce. Many components in the complex photosynthetic machinery of algal chloroplasts need replacing every few hours, yet the slugs maintain their captive collections of chloroplast hostages active for weeks or months until they die. Quite how they manage this feat is not clear, but it has led to the suspicion that the genes used by algae to run chloroplasts have somehow been transferred into the slugs' genome, the slugs having recruited them to build essential proteins for light harvesting and synthesizing chlorophyll (Rumpho, M.E. et al. (2008) Horizontal gene transfer of the algal nuclear gene psbO to the photosynthetic sea slug *Elysia chlorotica*. *Proceedings of the National Academy of Sciences, USA*, **105**, 17867–71). It's an exciting, if controversial, proposition for explaining the photosynthetic longevity of these startling lime-green creatures. The following paper, for example, reports evidence against the idea: Wägele, H. et al. (2011) Transcriptomic evidence that longevity of acquired plastids in the photosynthetic slugs *Elysia timida* and *Plakobranchus ocellatus* does not entail lateral transfer of algal nuclear genes. *Molecular Biology and Evolution*, **28**, 699–706. In contrast, this paper reports evidence supporting it: Schwartz, J.A., Curtis, N.E. & Pierce, S.K. (2010) Using algal transcriptome sequences to identify transferred genes in the sea slug, *Elysia chlorotica*. *Evolutionary Biology*, **37**, 29–37.
16. The highly controversial claim for evidence of fossil remains of a past Martian biota was originally reported here: McKay, D.S. et al. (1996) Search for past life on Mars: possible relic biogenic activity in Martian meteorite ALH84001. *Science*, **273**, 924–30.
17. Delwiche, C.F. & Timme, R.E. (2011) Plants. *Current Biology*, **21**, R417–22.
18. Delwiche & Cooper (2015).
19. Becker, B. & Marin, B. (2009) Streptophyte algae and the origin of embryophytes. *Annals of Botany*, **103**, 999–1004.
20. Harholt, A., Moestrup, P.O. & Ulvskov, P. (2016) Why plants were terrestrial from the beginning. *Trends in Plant Science*, **21**, 96–101. DeVries, J. & Archibald, J.M. (2018) Plant evolution: landmarks on the path to terrestrial life. *New Phytologist*, **217**, 1428–34.
21. See Delwiche & Cooper (2015) for a discussion.
22. Delwiche & Timme (2011).
23. Becker, B. (2013) Snowball earth and the split of Streptophyta and Chlorophyta. *Trends in Plant Science*, **18**, 180–3.
24. His argument rests on clues deciphered from the ancient metabolic pathways of algae. The first major Snowball Earth glaciation, known as the Sturtian glaciation, some 720

million years ago, was brought about by plummeting atmospheric levels of carbon dioxide that imposed conditions of carbon dioxide starvation on algae who struggled to photosynthesize. Chlorophytes living under dimly lit oceanic conditions beneath thick sea ice evolved energy conservation strategies for coping with carbon dioxide deprivation. Charophyte algae, on the other hand, living in temporary freshwater pools on top of the snowball, where sunlight was brighter, were not subject to the same energy constraints, and evolved energy-intensive metabolic strategies for survival.

25. Becker & Marin (2009).
26. Church, A.H. (1919) Thalassiophyta and the subaerial transmigration. *Oxford Botanical Memoirs*, **3**, 1–95.
27. Becker & Marin (2009).
28. Such conditions may even have prompted a primordial unicellular ancestor to undergo an evolutionary transition to a semi-terrestrial lifestyle that proved decisive in the origin of our land floras, see Harholt, A., Moestrup, P.O. & Ulvskov, P. (2016) Why plants were terrestrial from the beginning. *Trends in Plant Science*, **21**, 96–101.
29. Vermeij, G.J. & Dudley, R. (2000) Why are there so few evolutionary transitions between aquatic and terrestrial ecosystems? *Biological Journal of the Linnean Society*, **70**, 541–54.
30. Olsen, J.L. et al. (2016) The genome of the seagrass *Zostera marina* reveals angiosperm adaptation to the sea. *Nature*, **530**, 331–5.
31. Karol, K.G. et al. (2001) The closest living relatives of land plants. *Science*, **294**, 2351–3; Lewis, L.A. & McCourt, R.M. (2004) Green algae and the origin of land plants. *American Journal of Botany*, **91**, 1535–56.
32. Finet, C. et al. (2012) Multigene phylogeny of the green lineage reveals the origin and diversification of land plants. *Current Biology*, **22**, 1456–7. The authors use transcriptome data to derive phylogenetic relationships between Streptophytes and land plants. Nuclear DNA sequences of 77 genes from all lineages of Streptophytes and land plants produced a tree with Zygnematales and Coleochaetales as a sister group to land plants.
33. Wodniok, S. et al. (2011) Origin of land plants: do conjugating green algae hold the key? *BMC Evolutionary Biology*, **11**, 104). The authors used transcriptome data to derive phylogenetic relationships between Streptophytes and land plants. A nuclear sequence of 129 genes from all lineages of Streptophytes and land plants produced trees with either Zygnematales or Zygnematales + Coleochaetales as sister to land plants.
34. During sexual reproduction, each female sex organ (oogonium) contains one large, immobile egg, and each male sex organ (antheridium) produces one small, biflagellate sperm.
35. When crushed, these living relicts, the ancient relatives of which may have given rise to the terrestrial biosphere, release a strange garlic aroma. Only a few hundred species exist today, but they have a rich fossil record reaching back some 425 million years into the Silurian that documents their former glory.
36. Kelman, R. et al. (2004) Charophyte algae from the Rhynie chert. *Transactions of the Royal Society of Edinburgh: Earth Sciences*, **94**, 445–55.
37. McCourt, R.M., Delwiche, C.F. & Karol, K.G. (2004) Charophyte algae and land plant origins. *Trends in Ecology and Evolution*, **19**, 661–6.

38. Large-scale sequencing projects are now pointing to a new possibility—that the immediate ancestors of all land-dwellers belong to a clade of charophyte algae containing the Coleochaetales and the Zygnematales: Timme, R.E., Bachvaroff, T.R. & Delwiche, C.F. (2012) Broad phylogenomic sampling and the sister lineage of land plants. *PLoS One*, 7, e29696. In this study, the authors used transcriptome data to derive phylogenetic relationships between Streptophytes and land plants. They sampled the nuclear DNA sequences of 160 genes from all lineages of Streptophytes and land plants and the analyses resulted in trees with Zygnematales as sister to land plants. See also discussion by Bowman, J.L. (2013) Walkabout on the long branches of plant evolution. *Current Opinion in Plant Biology*, **16**, 70–7.
39. This view was championed by Linda Graham at the University of Wisconsin. See, for example, Graham, L.E. (1984) *Coleochaete* and the origin of land plants. *American Journal of Botany*, **71**, 603–8, and Graham, L.E. (1985) The origin of the life cycle of land plants. *American Scientist*, **73**, 178–86.
40. Graham, L.E. et al. (2012) Aeroterrestrial *Coleochaete* (Steptophyta, Coleochaetales) models early plant adaptation to land. *American Journal of Botany*, **99**, 130–44.
41. Niklas, K.J. (1976) Morphological and otogenetic reconstruction of *Parka decipiens* Fleming and *Pachytheca* Hooker from the Lower Old Red Sandstone, Scotland. *Transactions of the Royal Society of Edinburgh*, **69**, 483–99.
42. See, for examples, Shaw, A.J., Szovenyi, P. & Shaw, B. (2011) Bryophyte diversity and evolution: windows into early evolution of land plants. *American Journal of Botany*, **98**, 352–69.
43. Puttick, M.N. et al. (2018) The interrelationships of land plants and the nature of the ancestral embryophyte. *Current Biology*, **28**, 1–13. See also: Chang, C., Bowman, J.L. & Meyerowitz, E.M. (2016) A field guide to plant model systems. *Cell*, **167**, 325–39.
44. Wellman, C.H., Osterloff, P.L. & Mohiuddin, U. (2003) Fragments of the earliest land plants. *Nature*, **425**, 282–5.
45. Ligrone, R., Duckett, J.G. & Renzaglia K.S. (2012) Major transitions in the evolution of early land plants: a bryological perspective. *Annals of Botany*, **109**, 851–71.
46. Hernick, L.V., Landing, E. & Bartowski, K.E. (2008) Earth's oldest liverworts *Metzgeriothallus sharonae* sp. nov. from the Middle Devonian (Givetian) of eastern New York, USA. *Review of Palaeobotany and Palynology*, **148**, 154–62.
47. Guo, C-Q. et al. (2012) *Riccardiothallus devonicus* gen. et sp. nov., the earliest simple thalloid liverwort from the Lower Devonian of Yunnan, China. *Review of Palaeobotany and Palynology*, **176–7**, 35–40.
48. Heinrichs, J. et al. (2007) Evolution of two leafy liverworts: estimating divergence times from chloroplast DNA sequences using penalized likelihood with integrated fossil evidence. *Taxon*, **56**, 31–44. Crandall-Stotler, B., Stotler, R.E. & Long, D.G. (2009) Phylogeny and classifications of the Marchantiophyta. *Edinburgh Journal of Botany*, **66**, 155–98.
49. Some botanists have looked at the spores of *Haplomitrium* and found them to be very simple, with thin spore walls that presumably afforded limited protection from desiccation. Some have suggested that this limits the scope for their preservation, leading to the notion that the early land plant spore record is not preserved (Renzaglia,

K.S. et al. (2015) Permanent spore dyads are not 'a thing of the past': on their occurrence in the liverwort *Halpomitrium* (Haplomitriopsida). *Botanical Journal of the Linnean Society*, **179**, 658–69). But the fact is that spores tend to preserve, no matter how thin the wall. Sporopollenin is so robust that it always preserves. In fact, some of the early land-plant spores do have relatively thin walls compared to many modern spores and still preserve.

50. Edwards, D. et al. (2014) Cryptospores and cryptophytes reveal hidden diversity in early land floras. *New Phytologist*, **202**, 50–78.
51. Donoghue, P.C.J. & Yang, Z. (2016) The evolution of methods for establishing evolutionary timescales. *Philosophical Transactions of the Royal Society*, **B371**, 20160020.
52. Clarke, J.T. et al. (2011) Establishing a time-scale for plant evolution. *New Phytologist*, **192**, 266–301. See also the commentary: Kenrick, P. (2011) Timescales and timetrees. *New Phytologist*, **192**, 3–6. See also: Morris, J.L. et al. (2018) The timescale of early land plant evolution. *Proceedings of the National Academy of Science, USA*, **115(6)**, E2274–E2283.
53. Kenrick, P. et al. (2012) A timeline for terrestrialization: consequences for the carbon cycle of the Palaeozoic. *Philosophical Transactions of the Royal Society*, **B367**, 519–36.
54. Others have tried their hand at testing the early origin of land plants by crunching the DNA, dating the event to being close to the age of the oldest fossil spores (480–471 million years ago). See Magallón, S., Hilu, K.W. & Quandt, D. (2013) Land plant evolutionary timeline: gene effects are secondary to fossil constraints in relaxed clock estimation of age and substitution rates. *American Journal of Botany*, **100**, 556–73.
55. Rota-Stabelli, O., Daley, A.C. & Pisani, D. (2013) Molecular timetrees reveal a Cambrian colonization of land and a new scenario for ecdysozoan evolution. *Current Biology*, **23**, 392–8. See also the commentary: Dunn, C.W. (2013) Evolution: out of the ocean. *Current Biology*, **23**, R241. Lozano-Fernandez, J. et al. (2016) A molecular palaeobiological exploration of arthropod terrestrialization. *Philosophical Transactions of the Royal Society*, **B371**, 20150133.
56. Strother, P. et al. (2011) Earth's earliest non-marine eukaryotes. *Nature*, **473**, 505–9.
57. Controversial claims have been made that a distinctive geochemical signature in 850 million-year-old carbonate rocks lacked an obvious link to any sort of geological event. Instead, the explanation might be the activities of a photosynthetic terrestrial biosphere made up of algae, proto-land plants, and fungi; see Knauth, L.P. & Kennedy, M.J. (2009) The late Precambrian greening of the Earth. *Nature*, **460**, 728–32. For some issues with interpretation, see Arthur, M.A. (2009) Carbonate rocks deconstructed. *Nature*, **460**, 698–9.
58. Berner, R.A. (2004) *The Phanerozoic Carbon Cycle:* $CO_2$ *and* $O_2$. Oxford University Press, New York.
59. Kenrick, P. & Crane, P.R. (1997) The origin and early evolution of plants on land. *Nature*, **389**, 33–9.
60. Matsunaga, K.K.S. & Tomescu, A.M.F. (2016) Root evolution at the base of the lycophyte clade: insights from an Early Devonian lycophyte. *Annals of Botany*, **117**, 585–98.
61. DiMichele, W.A. & Phillips, T.L. (2002) The ecology of Paleozoic ferns. *Review of Palaeobotany and Palynology*, **119**, 143–59.

62. Delwiche & Timme (2011).
63. As discussed in relation to the oxygen-rich atmosphere of the time in Beerling, D.J. (2007) *The Emerald Planet. How plants changed Earth's history*. Oxford, Oxford University Press.
64. Nagalingum, N.S. et al. (2011) Recent synchronous radiation of a living fossil. *Science*, **334**, 796–9. See also the accompanying commentary: Renner, S.S. (2011) Living fossil younger than thought. *Science*, **334**, 766–7.
65. Butler, R.J. et al. (2009) Testing co-evolutionary hypotheses over geological timescales: interactions between Mesozoic non-avian dinosaurs and cycads. *Biological Reviews*, **84**, 73–89.
66. Herendeen, P.S. et al. (2017) Palaeobotanical redux: revisiting the age of the angiosperms. *Nature Plants*, **3**, doi:10.1038/nplants.2017.15.
67. Angiosperm Phylogeny Working Group III (2009) An update of the Angiosperm Phylogeny Group classification for the orders and families of flowering plants: APG III. *Botanical Journal of the Linnean Society*, **161**, 105–21. Chase, M.W. & Reveal, J.L. (2009) A phylogenetic classification of the land plants to accompany APG III. *Botanical Journal of the Linnean Society*, **161**, 122–7.
68. Around 97% of all living flowering plant species belong to one of these two major classes, but what of the fascinating remainder, the 3% that do not? They form a special group of plants that includes evolutionary lines with archaic features, a number of which arose before the great split between the monocots and the eudicots. Until recently, they had been lumped in with the monocots, but DNA evidence and fossils have since corrected the drawing of this part of the evolutionary tree. Probably the best known of these is a small collection of distinctive beautiful plants, the water lilies (Nymphaeales). Water lilies became adapted to a freshwater habitat during Cretaceous times and have remained in that watery milieu ever since. But the first of the surviving lineages to have split from the others is the family Amborellaceae, which consists of a single sickly-looking evergreen shrubby species restricted to the South Pacific Island of New Caledonia called *Amborella trichopoda*; of which more in Chapter Three. For now we should note *Amborella* is among the most primitive flowering plants alive today, with simple flowers. This fits together with evidence from one of the earliest flowering plants, a 120 million-year-old fossil plant called *Archaefructus*, which also has small unprepossessing flowers. So opinion over the past decade has shifted decisively towards the view that the original flowers possessed by early flowering plants were smaller and simpler than the exuberant examples we enjoy today. See Sun, G. et al. (1998) In search of the first flower: a Jurassic angiosperm, *Archaefructus*, from northeast China. *Science*, **282**, 1692–5. Note that subsequent radiometric dates placed *Archaefructus* in the Cretaceous rather than the Jurassic, see Sun, G. et al. (2002) Archaefructaceae, a new basal angiosperm family. *Science*, **296**, 899–904.
69. Updike, J. (2007) *Extreme dinosaurs*. National Geographic, December. http://ngm.nationalgeographic.com/print/2007/12/bizarre-dinosaurs/updike-text.
70. Niklas, K.J. & Kutschera, U. (2010) The evolution of the land plant life cycle. *New Phytologist*, **185**, 27–41.

71. Taylor, T.N., Kerp, H. & Hass, H. (2005) Life history biology of early land plants: deciphering the gametophyte phase. *Proceedings of the National Academy of Sciences, USA*, **102**, 5892–7. See also: Kenrick, P. (2017) How land plant life cycles first evolved. *Science*, **358**, 1538–9.
72. Ligrone, R., Duckett, J.G. & Renzaglia, K.S. (1993) The gametophyte–sporophyte junction in land plants. *Advances in Botanical Research*, **19**, 231–317.
73. Eisley, L. (1958) *The Immense Journey*. Lowe and Brydone, London.
74. Most researchers would accept that flowering plants evolved from wind-pollinated gymnosperm ancestors, but the nature of the fossil evidence makes it hard to offer definitive claims. For a discussion on the difficulties, see Friis, E.M., Crane, P.R. & Pedersen, K.R. (2011) *Early Flowers and Angiosperm Evolution*. Cambridge University Press, Cambridge.
75. We should be careful not to fall into the trap of imagining wind-pollination as a primitive condition. About 10% of today's flowering plant species rely on wind pollination, including those of grasslands and saltmarshes. As a method of reproduction, it evolved at least 65 separate times from animal-pollinated ancestor plant groups, raising the question: if it is such a wasteful mode of reproduction why has it evolved repeatedly in this manner? One answer seems to be that it evolves when animals become unreliable agents of pollen transfer; in this context wind pollination provides the reproductive assurance needed for setting seed. For discussion of these issues, see Barrett, S.C.H. (2010) Understanding plant reproductive diversity. *Philosophical Transactions of the Royal Society*, **B365**, 99–109.
76. These twin reproductive advantages follow from the process of double fertilization, a major contributing factor to the success of the angiosperms. Double fertilization takes place when a pollen grain extends into the flower, conveying two non-motile sperm cells to the ovule. One fertilizes the egg located inside the ovule, and the other initiates sexually formed embryo-nourishing tissue called endosperm. Cut open a maize seed and inside you'll find a visible illustration of the products of double fertilization. Pressed to one side is the plant embryo, the result of the first fertilization event, and surrounding it is the endosperm, a smooth substance filling the rest of the kernel, the product of the second fertilization event. Gymnosperm seeds enjoy no such benefits: only one of the two sperm cells conveyed by the pollen tube fertilizes the egg; the other degenerates. In a recent twist, William Friedman of Harvard University has shown that the gnetophytes have a rudimentary form of double fertilization that produces extra embryos. As we have already seen, gnetophytes (*Ephedra*, *Gnetum*, and *Welwitschia*) are gymnosperms closely related to flowering plants. It is easy to fall into the trap of thinking Friedman's findings suggest double fertilization already existed in some form in the common ancestor of flowering plants, but the molecular tree of life indicates this wasn't the case. The gnetophytes independently arrived at this innovative aspect of plant reproductive biology. For a review with historical context, see Friedman, W.E. (1998) The evolution of double fertilization and endosperm: an 'historical' perspective. *Sexual Plant Reproduction*, **11**, 6–16.
77. Herendeen et al. (2017).

78. Cardinal, S. & Danforth, B.N. (2013) Bees diversified in the age of eudicots. *Proceedings of the Royal Society*, **B280**, 20122686. Epiphytes, living on trunks and branches of trees, also seized the opportunities created by flowering plants, deriving moisture and nutrients from the rain and decomposing debris accumulating around them. Only DNA can tell us the story of epiphyte diversification, for these plants are rarely fossilized, and the spores they leave behind are often plain and unornamented, preserving no trace of their former diversity; see Schneider, H. et al. (2004) Ferns diversified in the shadow of angiosperms. *Nature*, **428**, 553–7.
79. Stromberg, C.A.E. (2011) Evolution of grasses and grassland ecosystems. *Annual Review of Earth and Planetary Sciences*, **39**, 517–44.
80. The metabolism of $C_4$ plants gives them the edge over their $C_3$ grassy cousins in warm, open, arid regions, and it is here where they tend to be most common, with a plentiful supply of sunlight powering their energy-demanding photosynthetic way of life. See Edwards, E.J. & Smith S.A. (2010) Phylogenetic analyses reveal the shady history of $C_4$ grasses. *Proceedings of the National Academy of Sciences, USA*, **107**, 2532–7.
81. Edwards, E.J. et al. (2010) The origin of $C_4$ grasslands: integrating evolutionary and ecosystem science. *Science*, **328**, 587–91.
82. Sage, R.F. et al. (2011) The $C_4$ plant lineages of planet Earth. *Journal of Experimental Botany*, **62**, 3155–69.
83. Sage, R.F. (2016) A portrait of the $C_4$ photosynthetic family on the 50th anniversary of its discovery: species number, evolutionary lineages, and Hall of Fame. *Journal of Experimental Botany*, **67**, 4039–56.
84. Spriggs, E.L., Christin, P.-A. & Edwards, E. J. (2014) $C_4$ photosynthesis promoted species diversification during the Miocene grassland expansion. *PLoS ONE*, **9 (5)**, e97722.
85. Arakaki, M. et al. (2011) Contemporaneous and recent radiations of the world's major succulent lineages. *Proceedings of the National Academy of Sciences, USA*, **108**, 8379–84.
86. Amis, M. (2002) *The War Against Cliché*. Vintage, London.

## 3. Genomes decoded

1. Miller, W. et al. (2008) Sequencing the nuclear genome of the extinct woolly mammoth. *Nature*, **456**, 387–90.
2. Green, R. E. et al. (2010) A draft sequence of the Neanderthal genome. *Science*, **328**, 710–22.
3. Frank, L. (2011) *My Beautiful Genome. Exposing Our Genetic Future, One Quirk at a Time*. One World, Oxford. Back in the days of the human genome project, whole genome sequencing required large teams of scientists and billions of dollars. Subsequent technological and computing advances have made it fast and cheap to sequence and analyse genomes. Many individuals have had their genomes sequenced for information regarding heritage as well as to obtain information regarding their likelihood of certain diseases. Before long, we may all have our genome sequences stored on our laptops or smart phones, and personalized approaches to medicine will be the 'new normal' for healthcare, Frank points out.

4. Hidalgo, O. et al. (2017) Is there an upper limit to genome size? *Trends in Plant Science*, 7, 567–73.
5. The Arabidopsis Genome Initiative (2000) Analysis of the genome sequence of the flowering plant *Arabidopsis thaliana*. *Nature*, **408**, 796–815.
6. Rensing, S.A. (2017) Why we need more non-seed plant models. *New Phytologist*, **216**, 355–60.
7. Derelle, E. et al. (2006) Genome analysis of the smallest free-living eukaryote *Ostreococcus tauri* unveils many unique features. *Proceedings of the National Academy of Sciences, USA*, **103**, 11647–52.
8. Archibald, J.M. (2006) Genome complexity in a lean, mean photosynthetic machine. *Proceedings of the National Academy of Sciences, USA*, **103**, 11433–4.
9. Worden, A.Z. et al. (2009) Green evolution and dynamic adaptations revealed by genomes of the marine picoeukaryotes *Micromonas*. *Science*, **324**, 268–72.
10. The larger genome of *Micromonas* relative to *Ostreococcus* includes a richer set of genes for transporting nutrients into the cell across cellular membranes, and for dealing with toxicity caused by heavy metals and reactive oxygen species. A clear advantage of the extended metabolic repertoire of *Micromonas* is that it copes with a wide range of environmental conditions, and this sees it occupy a global distribution throughout the world's oceans compared to the more restricted range of *Ostreococcus*; see Worden et al. (2009) above.
11. See Derelle et al. (2006).
12. See Derelle et al. (2006).
13. Merchant, S.S. et al. (2007) The *Chlamydomonas* genome reveals the evolution of key animal and plant functions. *Science*, **318**, 245–51.
14. Hodges, M.E. et al. (2012) The evolution of land plant cilia. *New Phytologist*, **195**, 526–40.
15. General review is given in Ainsworth, C. (2007) Tails of the unexpected. *Nature*, **448**, 638–41.
16. Li, J.B. et al. (2004) Comparative and basal genomics identifies a flagellar and basal body proteome that includes the *BBS5* human disease gene. *Cell*, **117**, 541–52.
17. See Knoll, A.H. (2011) The multiple origins of complex multicellularity. *Annual Reviews of Earth and Planetary Science*, **39**, 217–39.
18. Kirk, D.L. (1998) *Volvox. Molecular and genetic origins of multicellularity and cellular differentiation*. Cambridge University Press, Cambridge. See also: Kirk, D.L. (2010) *Volvox*. *Current Biology*, **14**, R599–600.
19. Multicellular *Volvox* is a physically larger organism than *Chlamydomonas*, but even so it is still dwarfed by the record-holding giant unicellular sub-tropical green alga by the name of *Acetabularia*, which is a whopping 20–40mm in diameter (Mandoli, D.F. (1998) Whatever happened to *Acetabularia*? Bringing the once classic model system into the age of molecular genetics. *International Reviews in Cytology*, **182**, 1–67).
20. Prochnik, S. E. et al. (2010) Genomic analysis of organismal complexity in the multicellular green alga *Volvox carteri*. *Science*, **329**, 223–6.
21. Nishii, I. & Miller, S.M. (2010) *Volvox*: simple steps to developmental complexity? *Current Opinion in Plant Biology*, **13**, 646–53.

22. Herron, M.D. et al. (2009) Triassic origin and early radiation of multicellular volvocine algae. *Proceedings of the National Academy of Sciences, USA*, **106**, 3254–8.
23. Stebbins, G.L. & Hill, G.C.J. (1980) Did multicellular plants invade the land? *American Naturalist*, **115**, 343. Harholt, J., Moestrup, O. & Ulvskov, P. (2016) Why plants were terrestrial from the beginning. *Trends in Plant Science*, **21**, 96–101. de Vries, J. et al. (2016) Streptophyte terrestrialization in light of plastid evolution. *Trends in Plant Science*, **21**, 467–76.
24. Hori, K. et al. (2014) *Klebsormidium flaccidum* genome reveals primary factors for plant terrestrial adaptation. *Nature Communications*, **5**, doi:10.1038/ncomms4978.
25. Nishiyama, T. et al. (2018) The *Chara* genome: secondary complexity and implications for plant terrestrialization. *Cell*, **174(2)**, 448–64.
26. Ju, C. et al. (2015) Conservation of ethylene as a plant hormone over 450 million years of evolution. *Nature Plants*, **1**, doi:10.1038/NPLANTS.2014.4.
27. Timme, R.E. & Delwiche, C.F. (2010) Uncovering the evolutionary origin of plant molecular processes: comparison of *Coleochaete* (Coleochaetales) and *Spirogyra* (Zygnematales) transcriptomes. *BMC Plant Biology*, **10**, 96.
28. Buschmann, H. & Zachgo, S. (2016) The evolution of cell division: from Streptophyte algae to land plants. *Trends in Plant Science*, **21**, 872–83.
29. Rensing, S.A. et al. (2008) The *Physcomitrella* genome reveals evolutionary insights in the conquest of land by plants. *Science*, **319**, 64–9.
30. Bowman, J.L. et al. (2017) Insights into land plant evolution garnered from the *Marchantia* polymorpha genome. *Cell*, **171**, 287–304. See also the commentary: Delwiche, C.F., Goodman, C.A. & Chang, C. (2017) Land plant model systems branch out. *Cell*, **171**, 265–6.
31. Chang, C., Bowman, J.L. & Meyerowitz, E.M. (2016) Field guide to plant model systems. *Cell*, **167**, 325–39.
32. Banks, J.A. et al. (2011) The *Selaginella* genome identifies genetic changes associated with the evolution of vascular land plants. *Science*, **332**, 960–2. For the evolutionary history of *Selaginella*, see Banks, J.A. (2009) *Selaginella* and 400 million years of separation. *Annual Reviews of Plant Biology*, **60**, 223–38.
33. Banks et al. (2011).
34. Chang, C., Bowman, J.L. & Meyerowitz, E. Field guide to plant model systems. *Cell*, **167**, 325–30.
35. Xu, B. et al. (2014) Contribution of NAC transcription factors to plant adaption to land. *Science*, **343**, 1505–8.
36. Robinson, J.M. (1990) Lignin, land plants, and fungi: biological evolution affecting Phanerozoic oxygen balance. *Geology*, **18**, 607–10.
37. Floudas, D. et al. (2012) Paleozoic origin of enzymatic lignin decomposition reconstructed from 31 fungal genomes. *Science*, **336**, 1715–19. See also the commentary by Hittinger, C.T. (2012) Endless rots most beautiful. *Science*, **336**, 1649–50.
38. Hibbett, D. et al. (2016) Climate, decay, and the death of coal forests. *Current Biology*, **26**, R563–7.
39. Nelson, M.P. et al. (2016) Delayed fungal evolution did not cause the Paleozoic peak in coal production. *Proceedings of the National Academy of Sciences, USA*, **113**, 2442–7.

Nevertheless, whatever the dating for white rot fungi, their arrival on the evolutionary stage must have changed the rates of terrestrial carbon cycle processes.

40. Weng, J.K. & Chapple, C. (2010) The origin and synthesis of lignin. *New Phytologist*, **187**, 273–85.
41. Niklas, K.J., Cobb, E.D. & Matas, J. (2017) The evolution of hydrophobic cell wall biopolymers: from algae to angiosperms. *Journal of Experimental Botany*, **68**, 5261–9.
42. Renault, H. et al. (2017) A phenol-enriched cuticle is ancestral to lignin evolution in land plants. *Nature Communications*, **8**, doi:10.1038/ncomms14713.
43. See discussion in Weng & Chapple (2010).
44. Emiliani, G. et al. (2009) A horizontal gene transfer at the origin of phenylpropanoid metabolism: a key adaptation of plants to land. *Biology Direct*, **4**, 7.
45. The current paradigm we have been rehearsing invokes lignin evolution in green plants as providing stems with mechanical rigidity for allowing shoots to grow upright. But very low levels of lignin were discovered in the cell walls of an intertidal red alga. We have already seen that red algae and vascular plants diverged over a billion years ago. One possibility is that the origin of the genes for lignin synthesis considerably pre-dated the origin of land plants. Another explanation is this represents another example of convergent evolution. Both may be a bit of a stretch. See Martone, P.T. et al. (2009) Discovery of lignin in seaweed reveals convergent evolution of cell-wall architecture. *Current Biology*, **19**, 169–75.
46. Weng, J.K. et al. (2008) Independent origins of syringyl lignin in vascular plants. *Proceedings of the National Academy of Sciences, USA*, **105**, 7887–92.
47. A comprehensive treatment of the topic is given in: Conway Morris, S. (2003) *Life's solution: Inevitable Humans in a Lonely Universe.* Cambridge University Press, Cambridge.
48. Melzer, S. et al. (2008) Flowering-time genes modulate meristem determinacy and growth form in *Arabidopsis thaliana*. *Nature Genetics*, **40**, 1489–92.
49. Kim, S.C. et al. (1996) A common origin for woody *Sonchus* and five related genera in the Macaronesian islands: molecular evidence for extensive radiations. *Proceedings of the National Academy of Sciences, USA*, **93**, 7743–8.
50. St Helena is one of the most desolate places on Earth. Lying 1200 miles south of the West African coast in the middle of the vast South Atlantic, it originally formed over a period of 7 million years about 14 million years ago. Eastwood, A., Gibby, M. & Cronk, Q.C.B. (2004) Evolution of St Helena arborescent Astereae (Asteraceae): relationships of the genera *Commidendrum* and *Melanodendron*. *Botanical Journal of the Linnean Society*, **144**, 69–83.
51. Crowther, T.W. et al. (2015) Mapping tree density at the global scale. *Nature*, **525**, 201–5.
52. Tuskan, G.A. et al. (2006) The genome of black cottonwood, *Populus tricharpa* (Torr. & Gray). *Science*, **313**, 1596–1604.
53. Myburg, A.A. et al. (2014) The genome of *Eucalyptus grandis*. *Nature*, **510**, 356–62.
54. See, for example, Lang, D. et al. (2010) Genome-wide phylogenetic comparative analysis of plant transcriptional regulation: a timeline of loss, gain, expansion, and correlation with complexity. *Genome Biology and Evolution*, **2**, 488–503. Wilhelmsson, P.K.I. et al. (2017) Comprehensive genome-wide classification reveals that many plant-specific transcription factors evolved in streptophyte algae. *Genome Biology and Evolution*, **9**, 3384–97.

55. Soltis, D.E., Visger, C.J. & Soltis, P.S. (2014) The polyploidy revolution then... and now: Stebbins revisited. *American Journal of Botany*, **101**, 1057–78.
56. Van de Peer, Y., Mizrachi, E. & Marchal, K. (2017) The evolutionary significance of polyploidy. *Nature Reviews Genetics*, **18**, 411–24.
57. Van de Peer, Y. (2004) Computational approaches to unveiling ancient genome duplications. *Nature Reviews Genetics*, **5**, 752–63.
58. Rensing, S.A. (2014) Gene duplication as a driver of plant morphogenetic evolution. *Current Opinion in Plant Biology*, **17**, 43–8.
59. Maere, S. et al. (2005) Modeling gene and genome duplications in eukaryotes. *Proceedings of the National Academy of Sciences, USA*, **102**, 5454–9.
60. Ibarra-Laclette, E. et al. (2013) Architecture and evolution of a minute plant genome. *Nature*, **498**, 94–8.
61. Guan, R. et al. (2016) Draft genome of the living fossil *Ginkgo biloba*. *GigaScience*, **5**, doi: 10.1186/s13742-016-0154-1.
62. Jiao, Y. et al. (2011) Ancestral polyploidy in seed plants and angiosperms. *Nature*, **473**, 97–100. See also the commentary by Van de Peer, Y. (2011) A mystery unveiled. *Genome Biology*, **12**, 113. Further evidence of ancient whole genome duplications in conifers is reported here: Li, Z. et al. (2015) Early genome duplications in conifers and other seed plants. *Science Advances*, **1**, e1501084.
63. Ruprecht, C. et al. (2017) Revisiting ancestral polyploidy in plants. *Science Advances*, **3**, e1603195.
64. Coen, E.S. & Meyerowitz, E.M. (1991) The war of the whorls: genetic interactions controlling flower development. *Nature*, **353**, 31–7.
65. For reviews, see Soltis, P.S. & Soltis, D.E. (2016) Ancient WGD events as drivers of key innovations in angiosperms. *Current Opinion on Plant Biology*, **30**, 159–65. Chanderbali, A.S. et al. (2016) Evolving ideas on the origin and evolution of flowers in the genomic era. *Genetics*, **202**, 1255–65.
66. Sanders, M.R. & Bowman, J.L. (2016) *Genetic Analysis: An Integrated Approach*. Global Edition. Pearson Education Limited, England.
67. Friis, E.M. et al. (2006) Cretaceous angiosperm flowers: innovations and evolution in plant reproduction. *Palaeogeography, Palaeoecology, Palaeoclimatology*, **232**, 251–93. See also: Herendeen, P.S. et al. (2017) Palaeobotanical redux: revisiting the age of the angiosperms. *Nature Plants*, **3**, doi: 10.1038/nplants.2017.15.
68. Van de Peer, Y. et al. (2009) The evolutionary significance of ancient genome duplications. *Nature Reviews Genetics*, **10**, 725–32.
69. Tank, D.C. et al. (2015) Progressive radiations and the pulse of angiosperm diversification. *New Phytologist*, **207**, 454–67.
70. Jao, Y. et al. (2014) Integrated syntenic and phylogenomic analyses reveal an ancient genome duplication in monocots. *Plant Cell*, **26**, 2792–802.
71. Estep, M.C. et al. (2014) Allopolyploidy, diversification, and the Miocene grassland expansion. *Proceedings of the National Academy of Sciences, USA*, **111**, 15149–54.
72. Salse, J. et al. (2008) Identification and characterization of shared duplications between rice and wheat provide new insight into grass genome evolution. *Plant Cell*, **20**, 11–24. See also Eckardt, N.A. (2008) Grass genome evolution. *Plant Cell*, **20**, 3–4.

73. See Haberer, G. et al. (2016) The big five of the monocot genomes. *Current Opinion in Plant Biology*, **30**, 33–40.
74. Edger, P.P. et al. (2015) The butterfly plant arms-race escalated by gene and genome duplications. *Proceedings of the National Academy of Sciences, USA*, **112**, 8362–6.
75. Ehrlich, P. & Raven, P.H. (1964) Butterflies and plants: a study in coevolution. *Evolution*, **18**, 586–608.
76. Ohno, S. (1970) *Evolution by Gene Duplication*. Springer-Verlag, New York.
77. Dehal, P. & Boore, J. L. (2005) Two rounds of whole genome duplication in the ancestral vertebrate. *PLoS Biology*, **3**, e314.
78. Van de Peer et al. (2009).
79. Zhou, X. et al. (2010) Phylogenetic detection of numerous gene duplications shared by animals, fungi and plants. *Genome Biology*, **11**, R38.
80. Beutler, E. (2002) Susumu Ohno 1928–2000. *Biographical Memoirs, National Academy of Sciences*, **81**, 1–13.
81. Fawcett, J.A. et al. (2009) Plants with double genomes might have had a better chance to survive the Cretaceous–Tertiary extinction event. *Proceedings of the National Academy of Sciences, USA*, **106**, 5737–42. See also the commentary by Soltis, D.E. & Burleigh, J.G. (2009) Surviving the K-T mass extinction: new perspectives of polyploidization in angiosperms. *Proceedings of the National Academy of Sciences, USA*, **106**, 5455–6.
82. Vanneste, K. et al. (2014) Analysis of 41 plant genomes supports a wave of successful genome duplications in association with the Cretaceous–Paleogene boundary. *Genome Research*, **24**, 1334–47. An update is provided by Lohaus, R. & Van de Peer, Y. (2016) Of dups and dinos: evolution at the K/Pg boundary. *Current Opinion in Plant Biology*, **30**, 62–9.
83. See Li et al. (2015).
84. As set out in my earlier book: Beerling, D.J. (2007) *The Emerald Planet. How Plants Changed Earth's History*. Oxford University Press, Oxford.

## 4. Ancient genes, new plants

1. Carpenter, H. (2000) *J.R.R. Tolkien: A Biography*. Houghton Mifflin, London.
2. Miller, A.J. et al. (2015) Expanding the role of botanical gardens in the future of food. *Nature Plants*, **1**, http://dx.doi.org/10.1038/nplants.2015.78
3. Kenrick, P. & Crane, P.R. (1997) The origin and early evolution of plants on land. *Nature*, **389**, 33–9; Bateman, R.M. et al. (1998) Early evolution of land plants: phylogeny, physiology, and ecology of the primary terrestrial radiation. *Annual Reviews in Ecology and Systematics*, **29** , 263–92.
4. See Kenrick, P. & Davis, P. (2004) *Fossil Plants*. The Natural History Museum, London.
5. Bateman et al. (1998).
6. See the following two helpful reviews. Dolan, L. (2009) Body building on land—morphological evolution of land plants. *Current Opinion in Plant Biology*, **12** , 4–8. Pires, N.D. & Dolan, L. (2012) Morphological evolution in land plants: new designs with old genes. *Philosophical Transactions of the Royal Society*, **B367**, 508–18.

7. Some are re-interpreting the fossil evidence to suggest miniature early vascular plants like *Cooksonia* were not hardy pioneers of the terrestrial landscape with adaptations later successfully exploited by vascular plants: see Boyce, C.K. (2008) How green was *Cooksonia*? The importance of size in understanding the early evolution of physiology in vascular land plants. *Paleobiology*, **34**, 174–94. Instead, Boyce (2008) suggests *Cooksonia* plants are simply structures for efficiently dispersing spores from miniature torpedo-shaped sacs at the tips of the stems. In this view, they are regarded as dependent—parasitic—upon that parental plant for nutrients and water, in an analogous manner to the sporophytes of mosses discussed in Chapter Two. Such arguments rest on size constraints. The very narrow axis diameters of some fossils—less than a millimetre—may be too small to house sufficient photosynthetic tissue to allow the plant to be free-living. Perhaps it was nourished by an unpreserved parental plant? Stimulating though these ideas are, they overlook the crucial benefit to photosynthesis and growth of very small plants of an atmosphere rich in carbon dioxide, with levels containing 10 or even 20 times as much as today's atmosphere.
8. Hetherington, A.J., Berry, C.M. & Dolan, L. (2016) Networks of highly branched stigmarian rootlets developed on the first giant trees. *Proceedings of the National Academy of Sciences, USA*, **113**, 6695–700.
9. I have previously discussed at length the argument that falling atmospheric carbon dioxide levels through the Devonian relieved an environmental constraint on the evolutionary origins and spread of flat-bladed (megaphyll) leaves. It is an example of how feedbacks between the slow geochemical cycling of carbon and plant activities constrained the evolution of plant form. See Beerling, D.J. (2007) *The Emerald Planet: How Plants Changed Earth 's History*. Oxford University Press, Oxford.
10. Hao, S.G., Beck, C.B. & Wang, D.M. (2003) Structure of the earliest leaves: adaptations to high concentrations of atmospheric $CO_2$. *International Journal of Plant Sciences*, **164**, 71–5.
11. Boyce, C.K. & Knoll, A.H. (2002) Evolution of developmental potential and the multiple independent origins of leaves in Paleozoic vascular plants. *Paleobiology*, **28**, 70–100. Boyce, C.K. (2010) The evolution of plant development in a palaeontological context. *Current Opinion in Plant Biology*, **13**, 102–7.
12. At least for seed-plant lineages we already have candidate gene families explaining the common evolutionary trajectories generating the diversity of leaf shapes: see Nardmann, J. & Werr, W. (2013) Symplesiomorphies in the *WUSCHEL* clade suggest that the last common ancestor of seed plants contained at least four independent stem cell niches. *New Phytologist*, **199**, 1081–92.
13. Weigel, D. & Jurgens, G. (2002) Stem cells that make stems. *Nature*, **415**, 751–4.
14. Sarkar, A.K. et al. (2007) Conserved factors regulate signalling in *Arabidopsis thaliana* shoot and root stem cell organizers. *Nature*, **446** , 811–14.
15. See Weigel & Jurgens (2002).
16. Kempin, S.A., Savidge, B. & Yanofsky, M.F. (1995) Molecular basis of the cauliflower phenotype in *Arabidopsis*. *Science*, **267**, 522–5.
17. See Weigel & Jurgens (2002).

18. Vollbrecht, E. et al. (1991) The developmental gene *Knotted-1* is a member of a maize homeobox gene family. *Nature*, **350**, 241–3.
19. Schneedberger, R. et al. (1998) The *rough sheath2* gene negatively regulates homeobox gene expression during maize leaf development. *Development*, **125**, 2857–65. Tsiantis, M. et al. (1999) The maize *rough sheath2* gene and leaf development programs in monocot and dicot plants. *Science*, **284**, 154–6.
20. Harrison, C.J. et al. (2005) Independent recruitment of a conserved developmental mechanism during leaf evolution. *Nature*, **434**, 509–14.
21. Byrne, M.E. et al. (2000) *Asymmetric leaves1* mediates leaf patterning and stem cell function in *Arabidopsis*. *Nature*, **408**, 967–71.
22. See Harrison et al. (2005).
23. Floyd, S.K. & Bowman, J.L. (2007) The ancestral developmental tool kit of land plants. *International Journal of Plant Science*, **168**, 1–35. Floyd, S.K. & Bowman, J.L. (2006) Distinct mechanisms reflect the independent origins of leaves in vascular plants. *Current Biology*, **16**, 1911–17. Floyd, S.K. & Bowman, J.L. (2010) Gene expression patterns in seed plant shoot meristems and leaves: homoplasy or homology? *Journal of Plant Research*, **123**, 43–55. Zalewski, C.S. et al. (2014) Evolution of the class IV *HD-Zip* gene family in streptophytes. *Molecular Biology and Evolution*, **30**, 2347–65.
24. Bower, F.O. (1935) *Primitive Land Plants*. Macmillan, London.
25. See Floyd & Bowman (2006) and also the interesting commentary on reconciliation of the differences of interpretation between Harrison et al. (2005) and Floyd & Bowman (2006): Kidner, C.A. (2007) Leaf evolution: working with what 's to hand. *Evolution and Development*, **9**, 321–2.
26. Raven, J.A. & Edwards, D. (2001) Roots: evolutionary origins and biogeochemical significance. *Journal of Experimental Botany*, **52**, 381–401. Algeo, T.J. & Scheckler, S.E. (1998) Terrestrial-marine teleconnections in the Devonian: links between the evolution of land plants, weathering processes and marine anoxic events. *Philosophical Transactions of the Royal Society*, **B353**, 113–30.
27. Kenrick, P. & Strullu-Derrien, C. (2014) The origin and early evolution of roots. *Plant Physiology*, **166**, 570–80.
28. Jones, V.A.S. & Dolan, L. (2012) The evolution of root hairs and rhizoids. *Annals of Botany*, **110**, 205–12.
29. Xue, J. et al. (2016) Belowground rhizomes in paleosols: the hidden half of an early Devonian vascular plant. *Proceedings of the National Academy of Sciences, USA*, **113**, 9451–6.
30. Kidston, R & Lang, W.H. (1920) On Old Red Sandstone plants showing structure, from the Rhynie Chert bed, Aberdeenshire. Part III *Asteroxylon mackiei*, Kidston and Lang. *Transactions of the Royal Society of Edinburgh*, **52**, 643–88. A recent assessment of where it fits into the modern framework of plant evolution is given in Kenrick, P.R. & Crane, P.R. (1997) *The Origin and Early Diversification of Land Plants: A Cladistic Study*. Smithsonian Institution Press, Washington, DC.
31. Hetherington, A.J., Dubrovsky, J.G. & Dolan, L. (2016) Unique cellular organization in the oldest root meristem. *Current Biology*, **26**, 1629–33.

32. Dolan, L. (2016) Q & A. *Current Biology*, **26**, R83–101.
33. Menand, B. et al. (2007) An ancient mechanism controls the development of cells with a rooting function in land plants. *Science*, **316**, 1477–80.
34. Menand et al. (2007).
35. Zimmer, A. et al. (2007) Dating the early evolution of plants: detection and molecular clock analyses of orthologs. *Molecular Genetics and Genomics*, **278**, 393–402.
36. Pires, N.D. et al. (2013) Recruitment and remodelling of an ancient gene regulatory network during land plant evolution. *Proceedings of the National Academy of Sciences, USA*, **110**, 9571–6.
37. Jang, G. et al. (2011) RSL genes are sufficient for rhizoid system development in early diverging plants. *Development*, **138**, 2273–81.
38. See Pires et al. (2013).
39. Lewis, D. (1983) Cyril Dean Darlington. 19 December 1903–26 March 1981. *Biographical Memoirs*, **29**, 113–57.
40. Harman, O.S. (2004) *The Man Who Invented the Chromosome: The Life of Cyril Darlington*. Harvard University Press, Cambridge, MA.
41. Willis, A.J. (1994) Arthur Roy Clapham. 24 May 1904–18 December 1990. *Biographical Memoirs*, **39**, 72–90. See also Pigott, D. (1992) Obituary: Arthur Roy Clapham, CBE, FRS (1904–1990). *Journal of Ecology*, **80**, 361–5.
42. Horton, P. (2012) David Alan Walker. 18 August 1928–13 February 2012. *Biographical Memoirs*, **60**, 413–32.
43. Proust, H. et al. (2016) RSL class I genes controlled the development of epidermal structures in the common ancestor of land plants. *Current Biology*, **26**, 93–9.
44. Bowman, J.L. et al. (2017) Insights into land plant evolution garnered from the *Marchantia polymorpha* genome. *Cell*, **171**, 287–304.
45. Puttick, M.N. et al. (2018) The interrelationships of land plants and the nature of the ancestral embryophyte. *Current Biology*, **28**, 1–13.
46. Bonnot, C. et al. (2017) Functional PTB phosphate transporters are present in streptophyte algae and early diverging land plants. *New Phytologist*, **214**, 1158–71.
47. Datta, S. et al. (2011) Roots hairs: development, growth and evolution at the plant–soil interface. *Plant Soil*, **346**, 1–14.
48. Dolan, L. (2017) Root hair development in grasses and cereals (Poaceae). *Current Opinion in Genetics and Development*, **45**, 76–81.
49. Wingen, H.R. et al. (2016) Mapping of quantitative trait loci for root hair length in wheat identifies loci that co-locate with loci for yield components. *Journal of Experimental Botany*, **67**, 4535–43.
50. Han, Y. et al. (2016) Altered expression of *TaRSL4* gene by genome interplay shapes root hair length in allopolyploid wheat. *New Phytologist*, **209**, 721–32.
51. See Kenrick, P. (2017) How land plant life cycles first evolved. *Science*, **358**, 1538–9.
52. Bowman, J.L. et al. (2016) Evolution in the cycles of life. *Annual Reviews of Genetics*, **50**, 133–54.
53. Master switches controlling such transitions are also involved. See for example, Horst, N.A. et al. (2016) A single homeobox gene triggers phase transition, embryogenesis

and asexual reproduction. *Nature Plants*, 2, article number 15209. See also the commentary: Hasebe, M. (2016) Starting BELL for embryos. *Nature Plants*, 2, article number 16004.

54. Sakakibara, K. et al. (2013) *KNOX2* genes regulate the haploid-to-diploid morphological transition in land plants. *Science*, **339**, 1067–70. See also the associated commentary article: Friedman, W.E. (2013) One genome, two ontogenies. *Science*, **339**, 1045–6.
55. Sano, R. et al. (2005) *KNOX* homeobox genes potentially have similar function in both diploid unicellular and multicellular meristems, but not haploid meristems. *Evolution and Development*, 7, 69–78.
56. Bowman et al. (2016).
57. For an accessible introduction into epigenetics, see Carey, N. (2012) *Epigenetics: How modern biology is rewriting our understanding of genetics, disease and inheritance*. Icon Books, London.
58. Originally discovered by research on fruit-flies (*Drosophila*), PcG may in fact be a universal protein complex present through the plant and animal kingdoms.
59. Mosquna, A. et al. (2009) Regulation of stem cell maintenance by the Polycomb protein FIE has been conserved during land plant evolution. *Development*, **136**, 2433–44.
60. Okano, Y. et al. (2009) A polycomb repressive complex 2 gene regulates apogamy and gives evolutionary insights into early land plant evolution. *Proceedings of the National Academy of Sciences, USA*, **106**, 16321–6.
61. Okano et al. (2009).
62. Mosquna et al. (2009).
63. For a dispatch from the frontline, see also Kenrick, P. (2017) How land plant life cycles first evolved. *Science*, **358**, 1538–9.
64. Went, F. (1935) Auxin, the plant growth hormone. *Botanical Review*, 1, 162–82. Not long after this triumph, Went left Utrecht to pursue his interests in plant science in Caltech in Pasadena, California, and went on to become the Director of Missouri Botanic Gardens in 1958, a position Peter Raven has held since 1971 (Chapter Eight); the web of such connections between the past and the present that reaches across the history of botanical science and discovery is built from such threads. For an accessible review of Darwin's work in a modern genomic context, see Holland, J.J. et al. (2009) Understanding phototropism: from Darwin to today. *Journal of Experimental Botany*, **60**, 1969–78.
65. Lau, S., Jurgens, G. & De Smet, I. (2008) The evolving complexity of the auxin pathway. *Plant Cell*, **20**, 1738–46. De Smet, I. et al. (2011) Unravelling the evolution of auxin signalling. *Plant Physiology*, **155**, 209–21.
66. Boot, K.J.M. et al. (2012) Polar auxin transport: an early invention. *Journal of Experimental Botany*, **63**, 4213–18. See also the extended commentary: Raven, J.A. (2013) Polar auxin transport in relation to long-distance transport of nutrients in the Charales. *Journal of Experimental Botany*, **64**, 1–9.
67. Viaene, T. et al. (2012) Origin and evolution of PIN auxin transporters in the green lineage. *Trends in Plant Science*, **18**, 5–10.

68. Viaene, T. et al. (2014) Directional auxin transport mechanisms in early diverging land plants. *Current Biology*, **24**, 1–6.
69. Sanders, H., Rothwell, G.W. & Wyatt, S.E. (2011) Parallel evolution of auxin regulation in rooting systems. *Plant Systematics and Evolution*, **291**, 221–5.
70. Yasumura, Y. et al. (2012) Studies of *Physcomitrella patens* reveal that ethylene-mediated submergence responses arose relatively early in land-plant evolution. *The Plant Journal*, **72**, 947–59.
71. Ju, C. et al. (2015) Conservation of ethylene as a plant hormone over 450 million years of evolution. *Nature Plants*, **1**, 1–7.
72. Harholt, J., Moestrup, O. & Ulvskov, P. (2016) Why plants were terrestrial from the beginning. *Trends in Plant Science*, **21**, 96–101.
73. The unfolding story of Harberd's discovery of the mechanisms by which DELLA proteins interact with gibberellin (GA) and the GID1 receptor to regulate plant growth is related in his engaging 2006 book *Seed to Seed: The Secret Life of Plants* (Bloomsbury, London).
74. Yabuta, T. & Sumiki, Y. (1938) On the crystal of gibberellin, a substance to promote plant growth. *Journal of the Agricultural Chemical Society of Japan*, **14**, 1526.
75. Stimulation of plant growth by GA can be demonstrated in genetically transformed *Arabidopsis* plants lacking a gene coding for an enzyme in its biosynthetic pathway. Transformed plants are effectively unable to synthesize their own gibberellin and develop a severely stunted appearance. Providing the plants with a source of GA corrects these developmental problems and identifies it as the causal agent responsible.
76. Navarro, L. et al. (2008) DELLAs control plant immune responses by modulating the balance of jasmonic acid and salicylic acid signaling. *Current Biology*, **18**, 650–5.
77. Three reviews capture the main details and timeline of the discoveries. Hirano, K. et al. (2008) GID1-mediated gibberellin signalling in plants. *Trends in Plant Science*, **13**, 192–9. Harberd, N.P., Belfield, E. & Yasumura, Y. (2009) The angiosperm gibberellin-GID1-DELLA regulatory mechanism: how an 'inhibitor of an inhibitor' enables flexible response to fluctuating environments. *The Plant Cell*, **21**, 1328–39. Sun, T. (2011) The molecular mechanism and evolution of the GA-GID1-DELLA signalling module in plants. *Current Biology*, **21**, R338–R345.
78. DELLA is an acronym derived from the five amino acid building blocks that stay the same in all members of the DELLA family of proteins. The five amino acids are aspartic acid (D), glutamic acid (E), leucine (L) (twice), and alanine (A). Amino acids have a one-letter code assigned to them, in a scheme created by pioneering bioinformatician Margaret Dayhoff (1925–1983), and are not always labelled by the first letter of their name to avoid repeats of the same letter. Single-letter abbreviations helped Dayhoff keep data-file sizes down in an era of punch-card computing.
79. Since the late 1950s, the presence of gibberellin has been reported in both seed plants and non-seed plants, including unicellular and multicellular algae, mosses, and ferns. Beyond implying great antiquity in the underlying genetic toolkits, nothing much else was known about what was going on. Yasumura, Y. et al. (2007) Step-by-step acquisition of the gibberellin-DELLA growth-regulatory mechanism during land plant evolution.

*Current Biology*, **17**, 1225–30. See also key research by Hirano, K. et al. (2007) The GID1-mediated gibberellin perception mechanism is conserved in the lycophyte *Selaginella moellendorfii* but not in the bryophyte *Physcomitrella patens*. *The Plant Cell*, **19**, 3058–79.

80. Achard, P. et al. (2008) Plant DELLAs restrain growth and promote survival of adversity by reducing the levels of reactive oxygen species. *Current Biology*, **18**, 656–60.
81. Peng, J. et al. (1999) 'Green revolution' genes encode mutant gibberellin response modulators. *Nature*, **400**, 256–61. Sasaki, A. et al. (2002) A mutant gibberellin-synthesis gene in rice. *Nature*, **416**, 701–2.
82. De Robertis, E.M. (2008) Evo-devo: variations on ancestral themes. *Cell*, **132**, 185–95. Carroll, S.B. (2008) Evo-devo and an expanding evolutionary synthesis: a genetic theory of morphological evolution. *Cell*, **134** , 25–36.
83. Monterio, A. & Podlaha, O. (2009) Wings, horns and butterfly eyespots: how do complex traits evolve? *PLoS Biology*, **7**, 209–16.
84. He was also writing before we realized that plants do rather better than the fourteenth-century alchemist Nicolas Flamel in postponing mortality. Flamel claimed to have succeeded in solving the twin magical puzzles of turning lead into gold and achieving immortality for his wife and himself. You have to question his conviction of his own immortality, though, when you discover he designed his own tombstone, which can still be found to this day preserved at the Musée de Cluny in Paris, the city where he lived into his 80s.
85. Schmid-Siegert, E., et al. (2017) Low number of fixed somatic mutations in a long-lived oak tree. *Nature Plants*, **3**, 926–9. See also the commentary: Kuhlemeier, C. (2017) How to get old without aging. *Nature Plants*, **3**, 916–17.
86. Larkin, P. (2011) *Poems. Selected and with an introduction by Martin Amis*. Faber and Faber Ltd, London.
87. See Niklas, K.J. & Kutschera, U. (2009) The evolutionary development of plant body plans. *Functional Plant Biology*, **36**, 682–95.

## 5. Gas valves

1. Graham, L.E. (1993) *Origin of Land Plants*. John Wiley and Sons, Inc., New York.
2. Edwards, D., Davies, K. & Axe, L. (1992) A vascular conducting strand in the early land plant *Cooksonia*. *Nature*, **357**, 683–5.
3. Edwards, D., Kerp, H. & Hass, H. (1998) Stomata in early land plants – an anatomical and ecophysiological approach. *Journal of Experimental Botany*, **49**, 255–78. Edwards, D. (1998) Climate signals in Palaeozoic land plants. *Philosophical Transactions of the Royal Society*, **B353**, 141–57.
4. Koch, G.W. et al. (2004) The limits to tree height. *Nature*, **428**, 851–4. See also the accompanying commentary: Woodward, F.I. (2004) Tall storeys. *Nature*, **428**, 807–8.
5. Dawson, T.E. (1998) Fog in the Californian redwood forest: ecosystem inputs and use by plants. *Oecologia*, **117**, 476–85.
6. See Dixon, H.H. & Joly, J. (1895) On the ascent of sap. *Philosophical Transactions of the Royal Society*, **B186**, 563–76. The mechanism by which water is raised against gravity to the upper foliage of tall trees puzzled plant physiologists for a long time. For a modern

perspective on the cohesion theory as an exclusive mechanism of long-distance water transport in plants, see Zimmermann, U. et al. (1994) Xylem water transport: is the available evidence consistent with the cohesion theory? *Plant, Cell & Environment*, **17**, 1169–81.

7. Nutrients synthesized by leaves flow in the opposite direction and in 1930 the German plant physiologist Ernst Münch (1876–1946) proposed an intuitive hypothesis to explain how. Nutrients should flow from areas with higher concentration (i.e., in the leaves, where sugars are synthesized and added to the system) to areas with lower concentration (i.e., the roots and fruits, where sugars are taken out). The flow of nutrients is passive and involves no expenditure of energy by the plant; an alternative would be a more complicated active system that uses energy to transport the nutrients through the tree. The Münch hypothesis has been described as 'super simple and super plausible', and is currently under detailed investigation in hard-to-access cells of tall trees.
8. Quote from Tyree, M.T. (2003) Plant hydraulics: The ascent of water. *Nature*, **423**, 923.
9. Beerling, D.J. (2015) Newton and ascent of water in plants. *Nature Plants*, **1**, 15005.
10. For balance, see also this critique: Tyree, M.T. (1997) The cohesion-tension theory of sap ascent: current controversies. *Journal of Experimental Botany*, **48**, 1753–65.
11. Conover, E. (2015) Gravity-defying trees explained by Newton. *Science*, February 2 2015. http://www.sciencemag.org/news/2015/02/gravity-defying-trees-explained-newton
12. Koch et al. (2004).
13. See Dawson (1998).
14. Berry, J.A., Beerling, D.J. & Franks, P.J. (2010) Stomata: key players in the Earth system, past and present. *Current Opinion in Plant Biology*, **13**, 233–40.
15. See Berry et al. (2010).
16. See Berry et al. (2010).
17. Hetherington, A.M. & Woodward, F.I. (2003) The role of stomata in sensing and driving environmental change. *Nature*, **424**, 903–8.
18. Choudhury, B. et al. (1998) A biophysical process-based estimate of global land surface evaporation using satellite and ancillary data—II. Regional and global patterns of seasonal and annual variations. *Journal of Hydrology*, **205**, 186–204.
19. Two papers provide the basis for this thinking. The first is a classic paper on the role of vegetation in influencing climate. It illustrates the role of plants in the current climate system and provides a good basis for speculating about the climate before the advent of vascular plants. See Shukla, J. & Mintz, Y. (1982) Influence of land-surface evapotranspiration on the earth's climate. *Science*, **215**, 1498–501. The second is a more recent example of the experiment above with a modern global climate and land surface model. It provides a good discussion of the role of vegetation in climate and how the climate system works. Read this if you want to go deeper than the Shukla and Mintz paper. Kleidon, A., Fraedrich, K. & Heimann, M. (2000) A green planet versus a desert world: estimating the maximum effect of vegetation on the land surface climate. *Climatic Change*, **44**, 471–93.
20. Dahl, E. et al. (2010) Devonian rise in atmospheric oxygen correlated to the radiations of terrestrial plants and large predatory fish. *Proceedings of the National Academy of*

*Sciences, USA*, **107**, 17911–15. Wallace, M.W. et al. (2017) Oxygenation history of the Neoproterozoic to early Phanerozoic and the rise of land plants. *Earth and Planetary Science Letters*, **466**, 12–19.

21. See Dahl et al. (2010). This fascinating idea has generated discussion. Butterfield, N.J. (2011) Was the Devonian radiation of large predatory fish a consequence of rising atmospheric oxygen concentration? *Proceedings of the National Academy of Sciences, USA*, **108**, E28. Dahl, E. et al. (2011) Reply to Butterfield: The Devonian radiation of large predatory fish coincided with elevated atmospheric oxygen levels. *Proceedings of the National Academy of Sciences, USA*, **108**, E29.
22. As discussed by Berry et al. (2010).
23. Sack, F.D. & Chen, J.-G. (2009) Pores in place. *Science*, **323**, 592–3.
24. Personal communication. Reflections on Berry's scientific career can be found in: Berry, J.A. (2012) There ought to be an equation for that. *Annual Reviews in Plant Biology*, **63**, 1–17.
25. A valuable source of historical information on stomatal research covering anatomy, physiology, and evolution is: Meidner, H. (1987) Three hundred years of research into stomata. In *Stomatal Function* (eds Zeiger, E., Farquhar, G.D., and Cowan, I.) pp. 7–27. Stanford University Press, Stanford.
26. Grew, N. (1682) *Anatomy of Plants*. W. Rawlins, London.
27. Bergmann is a protégé of Chris Somerville, who played a leading role in launching *Arabidopsis* on unsuspecting plant biologists. Prompted by James Watson, the Nobel Prize-winning co-discoverer of DNA, Somerville drove forward the sequencing of its compact genome. A historical perspective is provided by Somerville, C. & Koornneef, M. (2002) A fortunate choice: the history of *Arabidopsis* as a model plant. *Nature Reviews Genetics*, **3**, 883–9.
28. A profile of Torii is given in here: Torii, K.U. (2013) Q & A. *Current Biology*, **23**, R943–4.
29. Ohashi-Ito, K. & Bergmann, D.C. (2006) *Arabidopsis* FAMA controls the final proliferation/differentiation switch during stomatal development. *Plant Cell*, **18**, 2493–505.
30. MacAlister, C.A. et al. (2007) Transcription factor control of asymmetric cell divisions that establish stomatal lineages. *Nature*, **445**, 537–40. Pillitteri, L.J. et al. (2007) Termination of asymmetric cell division and differentiation of stomata. *Nature*, **445**, 501–5.
31. DeSmet, I. & Beeckman, T. (2011) Asymmetric cell division in land plants and algae: the driving force for differentiation. *Nature Reviews Molecular Cell Biology*, **12**, 177–87.
32. They belong to a family of proteins called basic-Helix-Loop-Helix (bHLH) transcription factors. These ancient proteins get their name from the fact that part of the chain of amino acids they code for contains a sequence of roughly 100 amino acids that are chemically attracted to each other in such a way that a characteristic loop forms as the chain folds around on itself. bHLH proteins are ubiquitous in eukaryotic organisms and likely evolved up to a billion years ago, before the divergence of the plant and animal kingdoms. Algal genomes have about five types, whereas the genomes of land plants, including mosses, lycophytes, and flowering plants, have over a hundred.

The diversity of bHLH proteins in green plants today appears to be a legacy of events over 400 million years ago when land plants originated and diversified. It is easy to imagine why. Expanding numbers of bHLH proteins could have orchestrated the evolutionary development of specialized cells and tissues required for adaptations to cope with the mounting environmental challenges faced by plants encountering harsh terrestrial environments (see Pires, N. & Dolan, L. (2009) Origin and diversification of basic-helix-loop-helix proteins in plants. *Molecular Biology and Evolution*, 27, 862–74).

33. MacAlister, C.A. & Bergmann, D.C. (2011) Sequence and function of basic helix-loop-helix proteins required for stomatal development in *Arabidopsis* are deeply conserved in land plants. *Evolution and Development*, **13**, 182–92.
34. The story is, as always, more complex than this. Mutant lines of *Arabidopsis* lacking *SPCH* were not rescued by the moss gene *SMF1*; for details see MacAlister & Bergmann (2011).
35. Kanaoka, M.M. et al. (2008) *SCREAM/ICE1* and *SCREAM2* specify three cell-state transitional steps leading to *Arabidopsis* stomatal differentiation. *Plant Cell*, **20**, 1775–85.
36. Chater, C.C. et al. (2016) Origin and function of stomata in the moss *Physcomitrella patens*. *Nature Plants*, 2, doi: 10.1038/nplants.2016.179.
37. Olsen, J. L. et al. (2016) The genome of the seagrass *Zostera marina* reveals angiosperm adaptation to the sea. *Nature*, **530**, 331–5.
38. Chater, C.C. et al. (2017) Origins and evolution of stomatal development. *Plant Physiology*, **174**, 624–38.
39. MacAlister et al. (2007).
40. Raissig, M.T. et al. (2016) Grasses use an alternatively wired bHLH transcription network to establish stomatal identity. *Proceedings of the National Academy of Sciences, USA*, **113**, 8326–31. Raissig, M.T. et al. (2017) Mobile MUTE specifies subsidiary cells to build physiologically improved grass stomata. *Science*, **355**, 1215–18.
41. Chen, Z.H. et al. (2017) Molecular evolution of grass stomata. *Trends in Plant Science*, **22**, 124–39.
42. Franks, P.J. & Farquhar, G.D. (2007) The mechanical diversity of stomata and its significance in gas-exchange control. *Plant Physiology*, **143**, 78–87.
43. Raissig et al. (2017).
44. Hetherington & Woodward (2003).
45. Dow, G.J., Bergmann, D.C. & Berry, J.A. (2014) An integrated model of stomatal development and leaf physiology. *New Phytologist*, **201**, 1218–26. Dow, G.J., Berry, J.A. & Bergmann, D.C. (2014) The physiological importance of developmental mechanisms that enforce proper stomatal spacing in *Arabidopsis thaliana*. *New Phytologist*, **201**, 1205–17. Dow, G.J. & Bergman, D.C. (2014) Patterning and processes: how stomatal development defines physiological potential. *Current Opinion in Plant Biology*, **21**, 67–74.
46. Rychel, A. L., Peterson, K. M. & Torii, K. U. (2010) Plant twitter: ligands under 140 amino acids enforcing stomatal patterning. *Journal of Plant Research*, **123**, 275–80.
47. Details have been reviewed and synthesized in Pillitteri, L.J. & Torri, K.U. (2012) Mechanisms of stomatal development. *Annual Review of Plant Biology*, **63**, 12.1–12.24.

48. Reviews of these processes are given in: Bergmann, D. C. & Sack, F. (2007) Stomatal development. *Annual Review in Plant Biology*, **58**, 163–81. Rowe, M. H. & Bergmann, D. C. (2010) Complex signals for simple cells: the expanding ranks of signals and receptors guiding stomatal development. *Current Opinion in Plant Biology*, **13**, 548–55.
49. Caine, R.S. et al. (2016) An ancestral stomatal patterning module revealed in the non-vascular land plant *Physcomitrella patens*. *Development*, **143**, 3306–14.
50. Davies, W.J. & Zhang, J.H. (1991) Root signals and the regulation of growth and development of plants in drying soils. *Annual Review of Plant Physiology and Plant Molecular Biology*, **42**, 55–76.
51. ABA was originally discovered in the 1960s by research groups investigating compounds that control bud dormancy and the shedding (abscission) of leaves and fruits. It is now recognized as an important hormone regulating many of the developmental and metabolic responses plants deploy when threatened by environmental dangers like a drying soil.
52. Root-to-shoot chemical communication of changes in water availability is of great commercial significance. Since the early 1990s, for example, Australian vineyards have exploited it by adopting a technique in which part of the vine's root system is allowed to dry while the other part is kept well-watered. Those roots experiencing the drying soil produce ABA signals that cause the stomatal pores on the leaves of the grapevines to partially close, reducing rates of water lost in transpiration. Partial-root-drying, as the technique is known, saves irrigation water and exerts no adverse effects on the quality of the grapes or wine, in some cases improving both. Saving water is important for the Australian viticulture industry as supplies for irrigation are an especially precious resource. Other major wine-making regions of the world, including South America, South Africa, California, and Southern Europe, are now trialling this promising approach to water conservation. A review in this area is given by Davies, W.J., Wilkinson, S. & Loveys, B. (2002) Stomatal control by chemical signalling and the exploitation of this mechanism to increase water use efficiency in agriculture. *New Phytologist*, **153**, 449–60.
53. Garner, D.L.B. & Paolillo, D.J.J. (1973) On the functioning of stomates in *Funaria*. *The Bryologist*, **76**, 423–7.
54. Chater, C. et al. (2011) Regulatory mechanism controlling stomatal behaviour conserved across 400 million years of land plant evolution. *Current Biology*, **21**, 1025–9. A commentary on this paper is given by Bowman (2011) Stomata: active portals for flourishing on land. *Current Biology*, **21**, R540–1.
55. Lind, C. et al. (2015) Stomatal guard cells co-opted ancient ABA-dependent desiccation survival system to regulate stomatal closure. *Current Biology*, **25**, 1–8. See also Cutler, S.R. et al. (2010) Abscisic acid: emergence of a core signalling network. *Annual Review of Plant Biology*, **61**, 651–79.
56. Discovery of *OST1* came when *Arabidopsis* mutants were subjected to a drought and screened with thermal imaging. Those plants with abnormally open stomata had cooler leaves than those without: see Mustilli, A.C. et al. (2002) *Arabidopsis* OST1 protein kinase mediates the regulation of stomatal aperture by abscisic acid and acts

upstream of reactive oxygen species production. *Plant Cell*, **14**, 3089–99. This screening technique would have appealed to Francis Darwin, who was captivated by the movement of guard cells over a century ago and a pioneer of a branch of botany called stomatal physiology, suitably encouraged and supported in his research at Cambridge University by A.C. Seward, whom we met in Chapter One. He recognized as early as 1904 that the transpirational flow of water through stomata cools plants, much as sweating cools our skin. Darwin, F. (1898) IX. Observations on stomata. *Philosophical Transactions of the Royal Society*, **B190**, 531–621.

57. Lind et al. (2015).
58. Julian Schroeder at the University of San Diego, and colleagues, published a colour heat map showing the similarity of all proteins involved in ABA perception, metabolism, and signalling in green plant functions relative to *Arabidopsis* (Hauser F., Waadt, R. & Schroeder, J.I. (2011) Evolution of abscisic acid synthesis and signalling mechanisms. *Current Biology*, **21**, R346–55). A similar exercise is performed here but with a greater emphasis on stomatal functioning: Cai, S. et al. (2017) Evolutionary conservation of ABA signaling for stomatal closure. *Plant Physiology*, **174**, 732–47.
59. Earlier suggestions that the ABA response evolved in lineages that evolved into ferns have proved unfounded and not stood the test of time. Brodribb, T.J. & McAdam, S.A.M. (2011) Passive origins of stomatal control in vascular plants. *Science*, **331**, 582–5. McAdam, S.A.M. & Brodribb, T.J. (2012) Fern and lycophyte guard cells do not respond to endogenous abscisic acid. *The Plant Cell*, **24**, 1510–21.
60. Investigations with the lycophyte *Selaginella* confirmed ABA induces a dose-dependent closure of stomata response. *Selaginella* employs a version of *OST1* that is closely related to that found in *Physcomitrella*; when substituted into *Arabidopsis* mutants lacking their own copy, it rescues their normal ABA-induced stomatal closing response. See Ruszala, E. et al. (2011) Land plants acquired active stomatal control early in their evolutionary history. *Current Biology*, **21**, 1030–5. Ferns also show an ABA stomatal closure response, see Cai et al. (2017) and also Hõrak, H., Kollist, H. & Merilo, E. (2017) Fern stomatal responses to ABA and $CO_2$ depend on species and growth conditions. *Plant Physiology*, **174**, 672–9.
61. Merced, A. & Renzaglia, K. (2014) Developmental changes in guard cell wall structure and pectin composition in the moss *Funaria*: implications for function and evolution of stomata. *Annals of Botany*, **114**, 1001–10.
62. Haig, D. (2013) Filial mistletoes: the functional morphology of moss sporophytes. *Annals of Botany*, **111**, 337–45.
63. Merced, A. & Renzaglia, K.S. (2013) Moss stomata in highly elaborated *Oedipodium* (Oedipodiaceae) and highly reduced *Ephemerum* (Pottiaceae) sporophytes are remarkably similar. *American Journal of Botany*, **100**, 2318–27.
64. See Chater et al. (2016).
65. See Keeley, J.E., Osmond, C.B. & Raven, J.A. (1984) *Stylites*, a vascular land plant without stomata, absorbs $CO_2$ via its roots. *Nature*, **310**, 694–5.
66. Beer, C. et al. (2010) Terrestrial gross carbon dioxide uptake: global distribution and covariation with climate. *Science*, **329**, 834–8.

67. McKown, A.D., Cochard, H. & Sack, L. (2010) Decoding leaf hydraulics with a spatially explicit model: principles of venation architecture and implications for its evolution. *American Naturalist*, **175**, 447–60.
68. Beerling, D.J. & Franks, P.J. (2010) The hidden cost of transpiration. *Nature*, **464**, 495–6.
69. Franks, P.J. & Beerling, D.J. (2009) Maximum leaf conductance driven by atmospheric $CO_2$ effects on stomatal size and density over geologic time. *Proceedings of the National Academy of Sciences, USA*, **106**, 10343–7.
70. Brodribb, T.J. et al. (2005) Leaf hydraulic capacity in ferns, conifers and angiosperms: impacts on photosynthetic maxima. *New Phytologist*, **165**, 839–46. Sack, L. & Holbrook, N.M. (2006) Leaf Hydraulics. *Annual Review of Plant Physiology and Molecular Biology*, **57**, 361–81.
71. For further discussion see also Brodribb, T.J. & Feild, T.S. (2009) Leaf hydraulic evolution led a surge in leaf photosynthetic capacity during early angiosperm diversification. *Ecology Letters*, **13**, 175–83.
72. Field, T.S. et al. (2011) Fossil evidence for Cretaceous escalation in angiosperm leaf vein evolution. *Proceedings of the National Academy of Sciences, USA*, **108**, 8363–6.
73. Betts, R.A. (1999) Self-beneficial effects of vegetation on climate in an ocean-atmosphere general circulation model. *Geophysical Research Letters*, **26**, 1457–60.
74. Boyce, C.K. & Lee, J-E. (2010) An exceptional role for flowering plant physiology in the expansion of tropical rainforests and biodiversity. *Proceedings of the Royal Society*, **B277**, 3437–43.
75. Spracklen, D.V., Arnold, S.R. & Taylor, C.M. (2012) Observations of increased tropical rainfall preceded by the passage of air currents. *Nature*, **489**, 282–5. See also the commentary by Arago, L.E.O.C. (2012) The rainforest's water pump. *Nature*, **489**, 217–18.
76. Further details of Heath's career are given in Mansfield, T.A. (1998) Oscar Victor Sayer Heath, 26 July 1903–16 June 1997. *Biographical Memoirs of Fellows of the Royal Society*, **44**, 219–35.
77. Heath, O.V.S. (1948) Control of stomatal movement by a reduction in the normal [$CO_2$] of the air. *Nature*, **161**, 179–81. Heath, O.V.S. & Russell, J. (1954) An investigation of the light responses of wheat stomata with the attempted elimination of control by the mesophyll. *Journal of Experimental Botany*, **5**, 1–15.
78. Keenan, T.F. et al. (2013) Increase in forest water-use efficiency as atmospheric carbon dioxide concentrations rise. *Nature*, **499**, 324–7.
79. Mansfield's obituary of O.V.S. Heath was written for *The Independent* newspaper (June 24, 1997).
80. Franks, P.J. et al. (2017) Stomatal function across temporal and spatial scales: deep-time trends, land-atmosphere coupling and global models. *Plant Physiology*, **174**, 583–602.
81. Sellers, P.J. et al. (1996) Comparison of radiative and physiological effects of doubled atmospheric $CO_2$ on climate. *Science*, **271**, 1402–6. A more recent analysis broadly confirming these effects is given by Cao, L. et al. (2010) Importance of carbon dioxide physiological forcing to future climate change. *Proceedings of the National Academy of Sciences, USA*, **107**, 9513–18.

82. The mammalian version was reported by Chandrashekar, J. et al. (2009) The taste of carbonation. *Science*, **326**, 443–5. The plant version was reported by Hu, H. et al. (2010) Carbonic anhydrases are upstream regulators of $CO_2$-controlled stomatal movements in guard cells. *Nature Structural Biology*, **12**, 87–93. An excellent commentary drawing these two accounts together is provided by Frommer, W.B. (2010) $CO_2$mmon sense. *Science*, **327**, 275–6.
83. See Hu et al. (2010).
84. Woodward, F.I. (1987) Stomatal numbers are sensitive to increases in $CO_2$ from preindustrial levels. *Nature*, **327**, 617–18.
85. The earliest reliable observations of atmospheric $CO_2$ were probably made on the coast of France from 1871 to 1880, see From, E. & Keeling, C.D. (1986) Reassessment of late 19th century atmospheric carbon dioxide variations in the air of western Europe and the British Isles based on an unpublished analysis of contemporary air masses by G. S. Callendar. *Tellus*, **38B**, 87–105.
86. Engineer, C.B. et al. (2014) Carbonic anhydrases, *EPF2* and a novel protease mediate $CO_2$ control of stomatal development. *Nature*, **513**, 246–50.
87. Gedney, N. et al. (2006) Detection of a direct carbon dioxide effect in continental river runoff records. *Nature*, **439**, 835–38.
88. Monteith, J.L. (1976) Closing remarks. *Philosophical Transactions of the Royal Society*, **B273**, 611–13.
89. Beerling, D.J. (2015) Gas valves, forests and global change: a commentary on Jarvis (1976) 'The interpretation of the variations in leaf water potential and stomatal conductance found in canopies in the field'. *Philosophical Transactions of the Royal Society*, **B370**, 20140311.

## 6. Ancestral alliances

1. Mackie, W. (1914) The rock series of Craigbed and Ord Hill, Rhynie, Aberdeenshire. *Transactions of the Edinburgh Geological Society*, **10**, 205–36.
2. In plural, lagerstätten. Lagerstätten have been discovered scattered across the globe and throughout the geological column, and collectively span a billion years of the history of life on Earth. The Burgess Shale lagerstätte in the Canadian Rocky Mountains dates to the early Cambrian, 505 million years ago, and ranks as one of the most famous for having yielded enormous insights into the sudden appearance of a wide diversity of complex animal life in the oceans.
3. See Edwards, D., Kenrick, P. & Dolan, L. (2018) History and contemporary significance of the Rhynie cherts – our earliest preserved terrestrial ecosystem. *Philosophical Transactions of the Royal Society*, **B373**, 20160489.
4. Mackie later had his 'if only' moment when realizing that the discovery of what is one of the most extraordinary sites of Devonian life yet unearthed could have been made 35 years earlier. If only he had followed up his chance discovery of fossil plant-bearing chert in the same region back in 1880. Historical details of the site's discovery are given in: Trewin, N.H. (2004) History of research on the geology and palaeontology of the

Rhynie area, Aberdeenshire, Scotland. *Transactions of the Royal Society of Edinburgh: Earth Sciences*, **94**, 285–98.

5. Kidston, R. & Lang, W.H. (1921) On Old Red Sandstone plants showing structure, from the Rhynie chert bed, Aberdeenshire. Part V. The thallophyta occurring in the peat-bed; the succession of the plants through a vertical section of the bed, and the conditions of accumulation and preservation of the deposits. *Transactions of the Royal Society of Edinburgh*, **52**, 855–902.
6. Thomson, C.A. & Wilkinson, I.P. (2009) Robert Kidston (1852–1924): biography of a Scottish geologist. *Scottish Journal of Geology*, **45**, 161–8. See also Lang, W.H. (1925) Robert Kidston—1852–1924. *Proceedings of the Royal Society*, **B98**, 14–22.
7. Pearson, H.L. (2014) Gender-bending in the Devonian at Rhynie and afterwards. *The Linnean*, **30**, 7–10; Pearson, H.L. (2016) Gender-bending in the Devonian at Rhynie: some corrections. *The Linnean*, **32**, 9–10.
8. Remy, W. et al. (1994) Four hundred-million-year-old vesicular arbuscular mycorrhizae. *Proceedings of the National Academy of Sciences, USA*, **91**, 11841–3.
9. Remy et al. (1994).
10. Taylor, T.N., Kerp, H. & Haas, H. (2005) Life history biology of early land plants: deciphering the gametophyte phase. *Proceedings of the National Academy of Sciences, USA*, **102**, 5892–7.
11. Dotzler, N. et al. (2006) Germination shields in Scuellospora (Glomeromycota: Diversisporales, Gigasporacea) from 400 million-year-old Rhynie chert. *Mycological Progress*, **5**, 178–84. Krings, M. et al. (2007) Fungal endophytes in a 400-million-yr-old land plant: infection pathways, spatial distribution, and host responses. *New Phytologist*, **174**, 648–57. See also the commentary by Berbee, M.L. & Taylor, J.W. (2007) Rhynie chert: a window into a lost world of complex plant–fungus interactions. *New Phytologist*, **174**, 475–9.
12. Pirozynski once helped Jane Goodall out at her camp in the Gombe Stream Reserve, East Africa, observing the chimpanzees that were to make her famous. Sitting in the dappled shade cast by the sparse canopies of savanna trees, he established the basis for our current estimates of global fungal biodiversity (Pirozynski, K.A. (1972) Microfungi of Tanzania. *Mycological Papers*, **129**, 1–64). Reasoning that fungal species outnumbered plant species by 3 to 1, possibly even 5 to 1 in the tropics, gave him a conservative estimate in the region of 1.5 million species, similar to the modern figure (Hawksworth, D.L. (2001) The magnitude of fungal diversity: the 1.5 million species estimate revisited. *Mycological Research*, **105**, 1422–32).
13. Pirozynski, K.A. & Malloch, D.W. (1975) The origin of land plants: a matter of mycotropism. *Biosystems*, **6**, 153–64. For the sake of historical accuracy, we should note that these authors, as they acknowledged, were following up on an idea sketched in outline only by Jeffrey, C. (1962) The origin and differentiation of the Archegoniate land plants. *Bot. Not.*, **115**, 446–54.
14. Ryan, F. (2003) *Darwin's Blind Spot: Evolution beyond Natural Selection*. Texere, Thompson Corporation, London. For detailed histories of the ideas related to symbiosis see Sapp, J. (1994) *Evolution by Association: A History of Symbiosis*. Oxford University Press, New York.

15. Smith, S.E. & Read, D.J. (2008) *Mycorrhizal Symbiosis* (3rd edn). Academic Press, London.
16. Tisserant, E. et al. (2013) Genome of an arbuscular mycorrhizal fungus provides insight into the oldest plant symbiosis. *Proceedings of the National Academy of Sciences,USA*, **110**, 20117–22.
17. For a review of reasoning for this date, see Edwards, D., Kenrick, P. & Dolan, L. (2018) History and contemporary significance of the Rhynie cherts – our earliest preserved terrestrial ecosystem. *Philosophical Transactions of the Royal Society*, **B373**, 20160489.
18. Zambonelli, A., Iotti, M. & Hall, I. (2015) Current status of truffle cultivation: recent results and future perspectives. *Micologia Italiana*, **44**, 31–40.
19. See Berbee, M.L. et al. (2017) Early diverging fungi: diversity and impact at the dawn of terrestrial life. *Annual Review of Microbiology*, **71**, 41–60. An earlier molecular clock dating attempt is given here: Simon, L. et al. (1993) Origin and diversification of endomycorrhizal fungi and coincidence with vascular land plants. *Nature*, **363**, 67–9.
20. Martin, F., Uroz, S. & Barker, D.G. (2017) Ancestral alliances: plant mutualistic symbioses with fungi and bacteria. *Science*, **356**, eaad4501.
21. Personal email communication to the author.
22. Margulis, L. (1998) *Symbiotic Planet: A New Look at Evolution*. Sciencewriters, Amherst, MA, USA.
23. See, for example, Ligrone, R. et al. (2007) Glomeromycotean associations in liverworts: a molecular, cellular and taxonomic analysis. *American Journal of Botany*, **94**, 1756–77.
24. Heinrichs, J. et al. (2007) Evolution of two leafy liverworts: estimating divergence times from chloroplast DNA sequences using penalized likelihood with integrated fossil evidence. *Taxon*, **56**, 31–44.
25. Krings, M. et al. (2007) An alternative mode of early land plant colonization by putative endomycorrhizal fungi. *Plant Signalling and Behaviour*, **2**, 125–6.
26. Note the liverworts possess single-celled rhizoids rather than true roots (Chapter Four), and this means that strictly speaking we cannot refer to their partnership with fungi as forming mycorrhiza. Instead, we should really use the awkward but more technically correct terminology of 'AM-like', but I have avoided this clumsy wording in the text.
27. Bowman, J.L. (2016) A brief history of *Marchantia* from Greece to genomics. *Plant and Cell Physiology*, **57**, 210–29.
28. Humphreys, C.P. et al. (2010) Mutualistic mycorrhiza-like symbiosis in the most ancient group of land plants. *Nature Communications*, **1**, 103, doi:10.1038/ncomms1105.
29. Bidartondo, M.I. et al. (2011) The dawn of symbiosis between plants and fungi. *Biology Letters*, **7**, 574–7.
30. Berbee, M.L., James, T.Y. & Strullu-Derrien, C. (2017) Early diverging fungi: diversity and impact at the dawn of terrestrial life. *Annual Review of Microbiology*, **71**, 41–60.
31. Puttick, M.N. et al. (2018) The interrelationships of land plants and the nature of the ancestral embryophyte. *Current Biology*, **28**, 1–13.
32. Carafa, A., Duckett, J.G. & Ligrone, R. (2003) Subterranean gametophyic axes in the primitive liverwort *Haplomitrium* harbour a unique type of endophytic association with aseptate fungi. *New Phytologist*, **160**, 185–97.
33. Duckett, J.G., Carafa, A. & Ligrone, R. (2006) A highly differentiated glomeromycotean association with the mucilage-secreting, primitive antipodean liverwort *Treubia* (Treubiaceae): clues to the origins of mycorrhizas. *American Journal of Botany*, **93**, 797–813.

Species of *Treubia* are copious secretors of mucilage. Production involves glandular structures that swell and rupture to discharge their contents. By absorbing water the mucilage can double in volume.

34. Field, K.J. et al. (2014) First evidence of mutualism between ancient plant lineages (Haplomitriopsida liverworts) and Mucoromycotina fungi and its response to simulated Palaeozoic changes in atmospheric $CO_2$. *New Phytologist*, **205**, 743–56.
35. Field et al. (2014).
36. Wang, B. et al. (2010) Presence of three mycorrhizal genes in the common ancestor of land plants suggests a key role of mycorrhizas in the colonization of land by plants. *New Phytologist*, **186**, 514–25. See also the accompanying commentary by Bonfante, P. & Selosse, M.A. (2010) A glimpse into the past of land plants and of their mycorrhizal affairs: from fossils to evo-devo. *New Phytologist*, **186**, 267–70.
37. For a review, see Strullu-Derrien, C. et al. (2018) The origin and evolution of mycorrhizal symbioses: from palaeomycology to phylogenomics. *New Phytologist*, doi: 10.1111/nph.15076.
38. Strullu-Derrien, C. et al. (2014) Fungal associations in *Horneophyton lignieri* from the Rhynie Chert (c. 407 million years old) closely resemble those in extant lower land plants: novel insights into ancestral plant–fungus symbioses. *New Phytologist*, **203**, 964–79. For a discussion of symbiotic strategies adopted by land plants during the 'greening of the land' see Field, K.J. et al. (2015) Symbiotic options for the conquest of the land. *Trends in Ecology and Evolution*, **30**, 477–486.
39. Wikström, N. & Kenrick, P. (2001) Evolution of the Lycopodiaceae (Lycopsida): estimating divergence times from rbcL gene sequences by use of nonparametric rate smoothing. *Molecular Phylogenetics and Evolution*, **19**, 177–86.
40. Lang, W.H. (1899) The prothallus of *Lycopodium clavatum*. *Annals of Botany*, **13**, 279–317. Lang, W.H. (1902) On the prothalli of *Ophioglossum pendulum* and *Helminthostachys zeylanica*. *Annals of Botany*, **16**, 23–62.
41. Kenrick, P. & Crane, P.R. (1997) The origin and early evolution of plants on land. *Nature*, **389**, 33–9.
42. Leake, J.R. (2005) Plants parasitic on fungi: unearthing the fungi in myco-heterotrophs and debunking the 'saprotrophic' plant myth. *The Mycologist*, **19**, 113–22.
43. Read, D.J. et al. (2000) Symbiotic fungal associations in 'lower' land plants. *Philosophical Transactions of the Royal Society*, **B355**, 815–31.
44. Leake, J.R. (1994) The biology of myco-heterotrophic ('saprophytic') plants. *New Phytologist*, **127**, 171–216. For an update, see Merckx, V., Bidartondo, M.I. & Hynson, N.A. (2009) Myco-heterotrophy: when fungi host plants. *Annals of Botany*, **104**, 1255–61.
45. Winther, J.L. & Friedman, W.E. (2007) Arbuscular mycorrhizal symbionts in *Botrychium* (Ophioglossaceae). *American Journal of Botany*, **94**, 1248–55. Winther, J.L. & Friedman, W.E. (2008) Arbuscular mycorrhizal associations in Lycopodiaceae. *New Phytologist*, **177**, 790–801. Winther, J.L. & Friedman, W.E. (2009) Phylogenetic affinity of arbuscular mycorrhizal symbionts in *Psilotum nudum*. *Journal of Plant Research*, **122**, 485–96. See also the commentary article drawing all this together by Leake et al. (2008).
46. As discussed by Leake, J.R., Cameron, D.D. & Beerling, D.J. (2008) Fungal fidelity in the myco-heterotroph-to-autotroph life cycle of Lycopodiaceae: a case of parental nurture. *New Phytologist*, **177**, 572–6.

47. This situation contrasts with the intergenerational transfer of carbon through the 'atmospheric commons' as a result of our combustion of fossil fuels, which is likely to have rather different consequences. Not only does it threaten the biodiversity of our land floras, as discussed in Chapter Eight, but it will also influence the climate of future generations not yet born, and who had no say in policies failing to regulate humanity's carbon emissions.
48. Leake et al. (2008).
49. Russell, A.F. & Hatchwell, B.J. (2001) Experimental evidence for kin-based helping in a cooperatively breeding vertebrate. *Proceedings of the Royal Society of London*, **B268**, 2169–74.
50. Cameron, D.D., Leake, J.R. & Read, D.J. (2006) Mutualistic mycorrhiza in orchids: evidence from plant–fungus carbon and nitrogen transfers in the green-leaved terrestrial orchid *Goodyera repens*. *New Phytologist*, **171**, 405–16.
51. Quote taken from Burkhardt, F. et al. (2003) *The Correspondence of Charles Darwin, Vol. II.* Cambridge University Press, Cambridge. It was recently unearthed and reported in a thorough review of Darwin's orchid research and its placement in a historical context by Yam, T.W., Arditti, J. & Cameron, K.M. (2009) 'The orchids have been a splendid sport' – an alternative look at Charles Darwin's contribution to orchid biology. *American Journal of Botany*, **96**, 2128–54.
52. Bernard, N. (1899) Sur la germination du *Neottia nidus-avis*. *Competes Rendus Académie des Sciences, Paris*, **128**, 1253–5.
53. Kiers, E.T. et al. (2011) Reciprocal rewards stabilize cooperation in the mycorrhizal symbiosis. *Science*, **333**, 880–2. See also the commentary: Selosse, M-A. & Rouseet, F. (2011) The plant-fungal market place. *Science*, **333**, 826–9.
54. Note that biologists are not always comfortable bringing economic theory into understanding symbioses between fungi and plants; for a discussion see Smith, F.A. & Smith, S. E. (2015) How harmonious are arbuscular mycorrhizal symbioses? Inconsistent concepts reflect different mindsets as well as results. *New Phytologist*, **205**, 1381–4.
55. Bidartondo, M.I. et al. (2002) Specialized cheating of the ectomycorrhizal symbiosis by an epiparasitic liverwort. *Proceedings of the Royal Society*, **B270**, 835–42.
56. Wickett, N.J. et al. (2007) Functional gene losses occur with minimal size reduction in the plastid genome of the parasitic liverwort *Aneura mirabilis*. *Molecular Biology and Evolution*, **25**, 393–401.
57. dePamphilis, C.W. & Palmer, J.D. (1990) Loss of photosynthetic and chlororespiratory genes from the plastic genome of a parasitic flowering plant. *Nature*, **348**, 337–9.
58. Plett, J.M. & Martin, F. (2011) Blurred boundaries: lifestyle lessons from ecomycorrhizal fungal genomes. *Trends in Genetics*, **27**, 14–22.
59. Young, N.D. et al. (2011) The *Medicago* genome provides insight into the evolution of rhizobial symbioses. *Nature*, **480**, 520–4.
60. Wang, B. et al. (2010) Presence of three mycorrhizal genes in the common ancestor of land plants suggests a key role of mycorrhizas in the colonization of land by plants. *New Phytologist*, **186**, 514–25. See also the accompanying commentary: Bonfante, P. & Selosse, M.A. (2010) A glimpse into the past of land plants and of their mycorrhizal affairs: from fossils to evo-devo. *New Phytologist*, **186**, 267–70.

61. Delaux, P.-M. et al. (2015) Algal ancestor of land plants was preadapted for symbiosis. *Proceedings of the National Academy of Sciences, USA*, **112**, 13390–5.
62. Stokstad, E. (2016) The nitrogen fix. *Science*, **353**, 1225–7.
63. Kloppholz, S., Kuhn, H. & Requena, N. (2011) A secreted fungal effector of *Glomus intraradices* promotes symbiotic biotrophy. *Current Biology*, **21**, 1204–9. See also the accompanying explanatory commentary by Sanders, I.R. (2011) Mycorrhizal symbioses: how to be seen as a good fungus. *Current Biology*, **21**, R550–2.
64. Terrer, C. et al. (2016) Mycorrhizal associations as a primary control of the $CO_2$ fertilization effect. *Science*, **353**, 72–3.
65. Salisbury, E.J. (1961) William Henry Lang. 1874–1960. *Biographical Memoirs of Fellows of the Royal Society*, **7**, 146–60.

## 7. Sculpting climate

1. A readable account of the story of fossil finds at Gilboa is given in VanAller Hernick, L. (2003) *The Gilboa Fossils*. New York State Museum, Albany, NY.
2. Goldring, W. (1927) The oldest known petrified forest. *Science Monthly*, **24**, 514–29.
3. Stein, W.E. et al. (1997) Giant cladoxylopsid trees resolve the enigma of the Earth's earliest forest stumps at Gilboa. *Nature*, **446**, 904–7. See also the accompanying commentary: Meyer-Berthaud, B. & Decombeix, A-L. (2007) Palaeobotany. A tree without leaves. *Nature*, **446**, 861–2.
4. Stein, W.E. et al. (2012) Surprisingly complex community discovered in the mid-Devonian fossil forest of Gilboa. *Nature*,**483**, 78–81. See also the accompanying commentary: Meyer-Berthaud, B. & Decombeix, A-L. (2012) Palaeobotany. In the shade of the earliest forest. *Nature*, **483**, 41–2.
5. Berner, R.A. (2013) From black mud to Earth system science: a scientific autobiography. *American Journal of Science*, **313**, 1–60.
6. Berner's thinking along these lines was set out in a couple of key papers. Berner, R.A. (1992) Weathering, plants, and the long-term carbon cycle. *Geochimica, Cosmochimica Acta*, **56**, 3225–31. Berner, R.A. (1997) The rise of plants and their effect on weathering and atmospheric $CO_2$. *Science*, **276** , 544–6.
7. Papers presented at the meeting were published in: Beerling, D.J., Chaloner, W.G. & Woodward, F.I. (eds) (1998) Vegetation–climate–atmosphere interactions: past, present and future. *Philosophical Transactions of the Royal Society*, **B353** , 1–171.
8. Basalt holds a special place in the slow dance of the geochemical carbon cycle. Despite the fact that it makes up less than 10% of the Earth's continental surface, it succumbs to chemical destruction by weathering relatively quickly. Iceland is ideal for these sorts of investigations because it is almost entirely composed of basalt, being located above the massive 65 000-km Mid-Atlantic Ridge that wraps around the floor of the Atlantic Ocean. Volcanic eruptions at this boundary create new ocean floor, inexorably forcing the North American and Eurasian tectonic plates apart at rates of 1 cm to 20 cm per year, a process known as sea-floor spreading. As oceanic plates move apart, molten rock wells up from tens of kilometres down, producing enormous volcanic eruptions of basalt.

9. Moulton, K.L. & Berner, R.A. (1998) Quantification of the effects of plants on weathering: studies in Iceland. *Geology*, **26**, 895–8. A detailed treatment is given in: Moulton, K.L., West, J. & Berner, R.A. (2000) Solute flux and mineral mass balance approaches to the quantification of plant effects on silicate weathering. *American Journal of Science*, **300**, 539–70.
10. See Moulton & Berner (1998) and Moulton et al. (2000).
11. These other studies were summarized by Berner et al. (2003) Phanerozoic atmospheric oxygen. *Annual Review of Earth and Planetary Sciences*, **31**, 105–34, and by Taylor, L.L. et al. (2009) Biological weathering and the long-term carbon cycle: integrating mycorrhizal evolution and function into the current paradigm. *Geobiology*, 7, 171–91.
12. Berner, R.A. (2004) *The Phanerozoic Carbon Cycle: $CO_2$ and $O_2$*. Oxford University Press, New York.
13. Mitchell, R.L. et al. (2016) Mineral weathering and soil development in the earliest land plant ecosystems. *Geology*, **44**, 1007–10.
14. Field evidence from boreal forests in north-western Ontario suggests lichens and mosses cause intense (but very shallow) chemical weathering leading to the production of secondary minerals, clays, and thin soils; these effects being absent from adjacent bare areas of the same granitic outcrop. See Jackson, T.A. (2015) Weathering, secondary mineral genesis, and soil formation caused by lichens and mosses growing on granitic gneiss in a boreal forest environment. *Geoderma*, **251–252**, 78–91.
15. Quirk, J. et al. (2015) Constraining the role of early land plants in Palaeozoic weathering and global cooling. *Proceedings of the Royal Society*, **B282**, 20151115, doi.org/10.1098/rspb.2015.1115.
16. Lenton, T.M. et al. (2012) First plants cooled the Ordovician. *Nature Geoscience*, **5**, 86–9.
17. Edwards, D., Cherns, L. & Raven, J.A. (2015) Could land-based early photosynthesizing ecosystems have bioengineered the planet in mid-Devonian times? *Palaeontology*, **58**, 803–37. See also experimental evidence with liverworts that when scaled up gives weathering fluxes 2–5% of contemporary trees: Quirk, J. et al. (2015) Constraining the role of early land plants in Palaeozoic weathering and global cooling. *Proceedings of the Royal Society*, **B282**, http://dx.doi.org/10.1098/rspb.2015.1115.
18. Chen, J., Blume, H.-P. & Beyer, L. (2000) Weathering of rocks induced by lichen colonization—a review. *Catena*, **39**, 121–46.
19. Mora, C.I., Driese, S.G. & Colarusso, L.A. (1996) Middle to late Paleozoic atmospheric $CO_2$ levels from soil carbonate and organic matter. *Science*, **271**, 1105–7.
20. The movement of Earth's tectonic plates, on which the continents sit, through the Devonian probably also played a role by slowly bringing landmasses into warmer, more humid low-latitude climate zones. Acting in concert with the spread of forests, this could have further hastened the weathering of silicate rocks and carbon dioxide sequestration into ocean sediments: see Hir, G.L. et al. (2011) The climate change caused by the land plant invasion in the Devonian. *Earth and Planetary Science Letters*, **310**, 203–12.
21. See, for example, Retallack, G.J. (1997) Early forest soils and their role in Devonian global change. Science, **276** , 583–5; Elick, J.M., Driese, S.G. & Mora, C.I. (1998) Very large plant and root traces from the Early to Middle Devonian: implications for early terrestrial ecosystems and atmospheric $p(CO_2)$. *Geology*, **26**, 143–6. An attempt at drawing

quantitative inferences from diverse literature reports regarding the evolutionary advance of trees and their effects on the global environment is reported here: Retallack, G.J. & Huang, C.M. (2011) Ecology and evolution of Devonian trees in New York, USA. *Palaeogeography, Palaeoclimatology, Palaeoecology*, **299**, 110–28.

22. Morris, J.L. et al. (2015) Investigating Devonian trees as geo-engineers of past climates: linking palaeosols to palaeobotany and experimental geobiology. *Palaeontology*, **58**, 787–801.
23. See Morris et al. (2015).
24. Morris, J. et al. (in prep) Early forest soils of the Middle Devonian, New York State, USA. *Palaios*; Stein, W.E. et al. (in prep) Mid Devonian root systems signal revolutionary change in earliest fossil forests. *Proceedings of the National Academy of Sciences, USA*.
25. See Goldring (1927).
26. Mahaffy, P.R. et al. (2015) The imprint of atmospheric evolution in the D/H of Hesperian clay minerals on Mars. *Science*, **347**, 412–4.
27. Kennedy, M. et al. (2006) Late Precambrian oxygenation: inception of the clay mineral factory. *Science*, **311**, 1446–9. See also the commentary article: Derry, LA. (2006) Fungi, weathering, and the emergence of animals. *Science*, **311**, 1386–7.
28. Hillier, S. (2006) Formation and alteration of clay materials. *Geological Society of London, Engineering Geology Special Publication*, **21**, 29–71.
29. See Kennedy et al. (2006).
30. Taylor, L.L. et al. (2009) Biological weathering and the long-term carbon cycle: integrating mycorrhizal evolution and function into the current paradigm. *Geobiology*, **7**, 171–91.
31. Berner, R.A. & Cochran, M.F. (1998) Plant-induced weathering of Hawaiian basalts. *Journal of Sedimentary Research*, **68**, 723–6.
32. Jongmans, A.G. et al. (1997) Rock-eating fungi. *Nature*, **389**, 682–3; Hoffland, E. et al. (2003) Feldspar tunnelling by fungi along natural productivity gradients. *Ecosystems*, **6**, 739–46.
33. Sverdrup, H. (2009) Chemical weathering of soil minerals and the role of biological processes. *Fungal Biology Reviews*, **23**, 94–100.
34. It gained the status of National Arboretum in 2001, the same year as Bedgebury Pinetum, Kent, was declared the country's National Pinetum. Bedgebury Pinetum is, as the name suggests, a complementary collection of conifers set in 350 acres of rolling Wealden countryside. It began life in the 1840s when established by the Beresford Hope family and suffered when the 1987 storm destroyed up to a third of the trees; it has since been extensively replanted.
35. Quirk, J. et al. (2012) Evolution of trees and mycorrhizal fungi intensifies silicate mineral weathering. *Biology Letters*, **8**, 1006–11.
36. Support for this view also turned up at the other end of the planet in the forests of South Island, New Zealand. Rock grains in soils beneath forests of southern beech (*Nothofagus menziesii*) and forests of Podocarpaceae trees are pockmarked with numerous small trenches, tunnels, and pits. These forests form partnerships exclusively with the recently evolved ectomycorrhizal fungi or the ancestral arbuscular mycorrhizal fungi, respectively, but the role of these fungi in making the tunnels in this study is unproven. In other words, field observations implicate

both groups of fungi in weathering processes. See: Koele, N. et al. (2014) Ecological significance of mineral weathering in ectomycorrhizal and arbuscular mycorrhizal ecosystems from a field-based comparison. *Soil Biology and Biogeochemistry*, **69**, 63–70.

37. Quirk, J. et al. (2014a) Ectomycorrhizal fungi and past high $CO_2$ atmospheres enhance mineral weathering through increased below-ground carbon-energy fluxes. *Biology Letters*, **10**, doi:10.1098/rsbl.2014.0375.
38. Howard, R.J. et al. (1991) Penetration of hard substrates by a fungus employing enormous turgor pressures. *Proceedings of the National Academy of Sciences, USA*, **88**, 11281–4.
39. Landeweert, R. et al. (2001) Linking plants to rocks: ectomycorrhizal fungi mobilize nutrients from minerals. *Trends in Ecology and Evolution*, **16**, 248–54. Van Scholl, L. et al. (2008) Rock-eating mycorrhizas: their role in plant nutrition and biogeochemical cycles. *Plant and Soil*, **303**, 35–47.
40. Leake, J.R. & Read, D.J. (2017) Mycorrhizal symbioses and pedogenosis throughout Earth's history. In: *Mycorrhizal Mediation of Soil: Fertility, Structure and Carbon Storage* (eds. Johnson, N.C., Gehring, C. & Jansa, J.), pp. 9–33, Elsevier, Amsterdam.
41. Jobbágy, E.G. & Jackson, R.B. (2001) The distribution of soil nutrients with depth: global patterns and the imprints of plants. *Biogeochemistry*, **53**, 51–77. Wardle, D.A., Walker, L.R. & Bardgett, R.D. (2004) Ecosystem properties and forest decline in contrasting long-term chronosequences. *Science*, **305**, 509–13.
42. Strullu-Derrien, C., Kenrick, P. & Selosse, M-A. (2017) Origin of the mycorrhizal symbioses. In *Molecular Mycorrhizal Symbiosis* (ed. Martin, F.), pp. 3–20, John Wiley & Sons, Hoboken, NJ, USA.
43. Quirk et al. (2014a).
44. See Leake & Read (2017).
45. Ordovician plant life, on the other hand, with minimal productivity and biomass and patchy cover, had limited scope for exerting such effects. Early land plants may have increased the respiratory generation of carbon dioxide and carbonic acid in shallow soils but fossil soils of that time suggest it had little effect on weathering effects prior to the evolution of vascular plants. See Jutras, P., LeForte, M.J. & Hanley, J.J. (2015) Record of climatic fluctuations and high pH weathering conditions in a thick Ordovician palaeosol developed in rhyolite of the Dunn Point Formation, Arisaig, Nova Scotia, Canada. *Geological Magazine*, **152**, 143–65.
46. Diaz, R.J. & Rosenberg, R. (2008) Spreading dead zones and consequences for marine ecosystems. *Science*, **321**, 926–9.
47. Cordell, D., Drangert, J.O. & White, S. (2009) The story of phosphorus: global food security and food for thought. *Global Environmental Change*, **19**, 262–305.
48. Bennett, E. & Elser, J. (2011) A broken biogeochemical cycle. *Nature*, **478**, 29–31.
49. See Gross, M. (2010) Fears over phosphorus supplies. *Current Biology*, **20**, R386–7.
50. Gosling, P. et al. (2006) Arbuscular mycorrhizal fungi and organic farming. *Agriculture Ecosystem and Environment*, **113**, 17–35.
51. See Bennett & Elser (2011).

52. For a discussion of the scientific history of these ideas, and the scientists involved, see Beerling, D.J. (2007) *The Emerald Planet: How Plants Changed Earth's History*. Oxford University Press, Oxford.
53. Pagani, M. et al. (2011) The role of carbon dioxide during the onset of Antarctic glaciation. *Science*, **334**, 1261–4.
54. Goddéris, Y. et al. (2017) Onset and ending of the late Palaeozoic ice age triggered by tectonically paced rock weathering. *Nature Geoscience*, **10**, 382–6.
55. Foster, G.L., Royer, D.L. & Lunt, D.J. (2017) Future climate forcing potentially without precedent in the last 420 million years. *Nature Communications*, **8**, doi: 10.1038/ncomms14845.
56. Detailed reconstructions of atmospheric carbon dioxide levels for Permo-Carboniferous glaciation, 300 million years ago, provide further confirmation of a lower bound of about two hundred parts per million, half what it is today: see Montañez, I.P. et al. (2016) Climate, $pCO_2$ and terrestrial carbon cycle linkages during late Palaeozoic glacial-interglacial cycles. *Nature Geoscience*, **9**, 824–8.
57. Pagani, M. et al. (2009) The role of terrestrial plants in limiting atmospheric $CO_2$ decline over the past 24 million years. *Nature*, **460**, 85–8. See also the accompanying commentary: Goddéris, Y. & Donnadieu, Y. (2009) Climatic plant power. *Nature*, **460**, 40–1.
58. Bond, W.J. & Midgley, G.F. (2012) Carbon dioxide and the uneasy interactions of trees and savannah grasses. *Philosophical Transactions of the Royal Society*, **B367**, 601–12. For low carbon dioxide effects, see Quirk, J. et al. (2014b) Weathering by tree root-associating fungi diminishes under simulated Cenozoic atmospheric $CO_2$ decline. *Biogeosciences*, **11**, 321–31.
59. Quirk, J. et al. (2013) Increased susceptibility to drought-induced mortality in *Sequoia sempervirens* (Cupressaceae) trees under Cenozoic atmospheric carbon dioxide starvation. *American Journal of Botany*, **100**, 582–91.
60. Gislason, S.R. et al. (2009) Direct evidence of the feedback between climate and weathering. *Earth and Planetary Science Letters*, **277**, 213–22.
61. Le Quéré, C. et al. (2016) Global carbon budget 2016. *Earth System Science Data*, **8**, 605–49.
62. Hansen, J. et al. (2017) Young people's burden: requirement of negative $CO_2$ emissions. *Earth Systems Dynamics*, **8**, 577–616.
63. Paris Agreement 2015, UNFCCC secretariat, available at https://unfccc.int/process/the-paris-agreement/what-is-the-paris-agreement (last accessed: 21 April 2017).
64. Smith, P. et al. (2016) Biophysical and economic limits to negative $CO_2$ emissions. *Nature Climate Change*, **6**, 42–50.
65. Schrag, D.P. (2009) Storage of carbon dioxide in offshore sediments. *Science*, **325**, 1658–9.
66. http://www.carbfix.com
67. Matter, J.M. et al. (2016) Rapid mineralization for permanent disposal of anthropogenic carbon dioxide emissions. *Science*, **352**, 1312–14.

68. The Columbia River basalts formed during a major episode of volcanic activity around 16 million years ago. Highly porous rock formed when bubbles of carbon dioxide migrated through the magma as it cooled. In this region, the porous basalt rock is sandwiched between layers of solid rock that prevent leaking and trap the dissolved carbon dioxide while it reacts to form carbonates: see Tollefson, J. (2013) Pilot projects bury carbon dioxide in basalt. *Nature*, **500**, 18.
69. Hansen et al. (2017).
70. Anderson, K. (2015) Talks in the city of light generate more heat. *Nature*, **528**, 437.
71. Beerling, D.J. et al. (2018) Farming with crops and rocks to address global climate, food and soil security. *Nature Plants*, **4**, 138–47.
72. For a collection of papers addressing this topic, see Beerling, D.J. (2017) (ed.) Enhanced rock weathering: biological climate change mitigation with co-benefits for food security. Mini-Series. *Biology Letters*, **13**, Issue 4.
73. Hoegh-Guldberg, O. et al. (2015) *Reviving the Ocean Economy: The Case for Action 2015*. World Wide Fund for Nature, Switzerland.
74. Renforth, P. et al. (2011) Silicate production and availability for mineral carbonation. *Environmental Science and Technology*, **45**, 2035–41.
75. Tubana, B.S., Babu, T. & Datnoff, L.E. (2016) A review of silicon in soils and plants and its role in US agriculture: history and future perspectives. *Soil Science*, **181**, 393–411.
76. Godfray, H.J.C. et al. (2010) Food security: The challenge of feeding 9 billion people. *Science*, **327**, 812–18.


## 8. Eden under siege


1. Butchart, S.H.M. et al. (2010) Global biodiversity: indicators of recent declines. *Science*, **328**, 1164–8.
2. The Missouri Botanic was founded by the philanthropist Henry Shaw (1800–1889). Sheffield-born Shaw made his fortune selling steel from Sheffield ('the Steel City') to pioneer settlers in the then small French town of St Louis on the banks of the Mississippi River. Within two decades, he was able to retire at the young age of 40 and spend his retirement working with botanists to plan, fund, and build the historic Missouri Botanical Garden. The Sheffield and the Missouri Botanical Gardens were forged in the crucible of the Industrial Revolution by co-operation between human societies, when the population was less than two billion, when the atmospheric carbon dioxide concentration was 30% lower than today, and biodiversity had not yet flinched.
3. Anon (2017) Twenty-first century botany. *Nature Plants*, **3**, 681.
4. May, R.M. (1978) Human reproduction reconsidered. *Nature*, **272**, 491–5.
5. Smil, V. (1999) Detonator of the population explosion. *Nature*, **400**, 415.
6. United Nations (2011) *World Population Prospects, the 2010 Revision. Volume 1. Comprehensive Tables*. Department of Economic and Social Affairs, New York. See also reviews: Bloom, D.E. (2011) 7 billion and counting. *Science*, **333**, 562–8; Lee, R. (2011) The outlook for population growth. *Science*, **333**, 569–73.

7. Gerland, P. et al. (2014) World population stabilization unlikely this century. *Science*, **346**, 234–7.
8. Godfrey, H.C.J. et al. (2010) Food security: the challenge of feeding 9 billion people. *Science*, **327**, 812–18.
9. Rands, M.R.W. et al. (2010) Biodiversity conservation: challenges beyond 2010. *Science*, **329**, 1298–1303.
10. Foley, J.A. et al. (2011) Solutions for a cultivated planet. *Nature*, **478**, 337–42.
11. Watson, E.M. et al. (2016) Catastrophic declines in wilderness areas undermine global environment targets. *Current Biology*, **26**, 2929–34.
12. Maxwell, S. et al. (2016) The ravages of guns, nets and bulldozers. *Nature*, **536**, 143–5.
13. Le Quéré, C. et al. (2016) Global carbon budget 2016. *Earth Systems Science Data*, **8**, 605–49.
14. Beerling, D.J. & Royer, D.L. (2011) Convergent Cenozoic $CO_2$ history. *Nature Geoscience*, **4**, 418–20.
15. Hansen, J. et al. (2013) Assessing 'Dangerous Climate Change': Required reduction of carbon emissions to protect young people, future generations and nature. *PLoS One*, **8**, e81648.
16. An excellent review of the Anthropocene concept is given in the following paper, whose authors include its originator, the Nobel Prize winner Paul Crutzen: Steffen, W. et al. (2011) The Anthropocene: conceptual and historical perspectives. *Philosophical Transactions of the Royal Society*, **A369**, 842–67.
17. Waters, C.N. et al. (2016) The Anthropocene is functionally and stratigraphically distinct from the Holocene. *Science*, **351**, 137–48.
18. http://science.kew.org/strategic-output/state-worlds-plants
19. Royal Botanic Gardens Kew (2010) *Plants under Pressure: A Global Assessment. The First Report of the IUCN Sampled Red List Index for Plants*. RBG, Kew.
20. Pimm, S.L. et al. (2014) The biodiversity of species and their rates of extinction, distribution, and protection. *Science*, **344**, doi: 10.1126/science.1246752.
21. Goettsch, B. et al. (2015) High proportion of cactus species threatened with extinction. *Nature Plants*, **1**, article number 15142.
22. Pitman, N.C.A. & Jorgensen, P.M. (2002) Estimating the size of the world's threatened flora. *Science*, **298**, 989.
23. Firbeck is noted for its oval green that was once the private racecourse of the eighteenth-century racehorse owner Anthony St Leger, the man who established the St. Leger Stakes horse race.
24. A modern context of Watson's work is provided in: Hubbell, S.P. (2001) *The Unified Neutral Theory of Biodiversity and Biogeography*. Princeton University Press, Princeton, NJ.
25. Myers, N. et al. (2000) Biodiversity hotspots for conservation priorities. *Nature*, **403**, 853–8.
26. Pimm, S.L. & Raven, P. (2000) Extinction by numbers. *Nature*, **403**, 843–5.
27. Thomas, C.D. et al. (2004) Extinction risk from climate change. *Nature*, **427**, 145–8. See also the accompanying commentary: Pounds, J.A. & Puschendorf, R. (2004) Clouded futures. *Nature*, **427**, 107–9.
28. Hubbell, S.P. et al. (2008) How many tree species are there in the Amazon and how many of them will go extinct? *Proceedings of the National Academy of Sciences, USA*, **105** **(Suppl 1)**, 11498–504.

29. Peres, C.A. et al. (2016) Dispersal limitation induces long-term biomass collapse in overhunted Amazonian forests. *Proceedings of the National Academy of Sciences, USA*, **113**, 892–7.
30. Wearn, O.R., Reuman, D.C. & Ewers, R.M. (2012) Extinction debt and the conservation opportunity in the Brazilian Amazon. *Science*, **337**, 228–32.
31. Pimm, S.L. (2008) Biodiversity: climate change or habitat loss – which will kill more species? *Current Biology*, **18**, R117–19.
32. Thuiller, W. et al. (2005) Climate change threats to plant diversity in Europe. *Proceedings of the National Academy of Sciences, USA*, **102**, 8245–50.
33. Lenoir, J. et al. (2008) A significant upward shift in plant species optimum elevation during the 20th century. *Science*, **320**, 1768–71. Dainese, M. et al. (2017) Human disturbance and upward expansion of plants in a warming climate. *Nature Climate Change*, 7, 577–80.
34. Steinbauer, M.J. et al. (2018) Accelerated increase in plant species richness on mountain summits is linked to warming. *Nature*, **556**, 231–4.
35. One issue is how well the species–area relationship describes the magnitude of extinctions as a result of habitat loss. Consider that, with increasing area, the species–area relationship increases each time the first individual of a new species is encountered. Further individuals of the same species add nothing to the species count, of course. Now, consider the reverse situation. Decreasing habitat area doesn't actually cause extinction until the last individual of that species is eliminated. It can be shown mathematically that predicting the extent of species extinctions using species–area relationships by moving down the slope overestimates the magnitude of species extinction. This is not to say we should be complacent about extinctions due to habitat destruction; rather that the estimates of Thomas et al. (2004) are probably towards the high end. See He, F. & Hubbell, S.P. (2011) Species–area relationships always overestimate extinction rates from habitat loss. *Nature*, **473**, 368–71, with an accompanying explanatory commentary by Rahbek, C. & Colwell, R.K. (2011) Species loss revisited. *Nature*, **473**, 288–9.
36. Woodward, F.I. & Kelly, C.K. (2008) Responses of global plant diversity to changes in carbon dioxide concentration and climate. *Ecology Letters*, **11**, 1229–37.
37. Velland, M. et al. (2017) Plant biodiversity change across scales during the Anthropocene. *Annual Reviews in Plant Biology*, **68**, 563–86.
38. Loarie, S.R. et al. (2009) The velocity of climate change. *Nature*, **462**, 1052–5. Further analyses are given in Diffenbaugh, N.S. & Field, C.B. (2013) Changes in ecologically critical terrestrial climate conditions. *Science*, **341**, 486–92.
39. Williams, J.W., Jackson, S.T. & Kutzbach, J.E. (2007) Projected distributions of novel and disappearing climates by 2100AD. *Proceedings of the National Academy of Sciences, USA*, **104**, 5738–42.
40. Tilman, D. et al. (1994) Habitat destruction and the extinction debt. *Nature*, **371**, 65–6.
41. Janzen, D.H. (2001) *Encyclopaedia of Biodiversity*, **4**, 590.
42. Cronk, Q.C.B. (1995) *The Endemic Flora of St Helena*. Anthony Nelson Ltd, Oswestry.
43. Cronk, Q. (2016) Plant extinctions take time. *Science*, **353**, 446–7.
44. Pimm et al. (2014). See also Velland et al. (2017).

45. Barnosky, A.D. et al. (2011) Has the Earth's sixth mass extinction already arrived? *Nature*, **471**, 51–7.
46. See Barnosky et al. (2011).
47. Benton, M.J. & Twitchett, R.J. (2003) How to kill (almost) all life: the end-Permian extinction event. *Trends in Ecology and Evolution*, **18**, 358–65.
48. Hautmann, M., Benton, M.J. & Tomašových, A. (2008) Catastrophic ocean acidification at the Triassic–Jurassic boundary. *Neues Jahrbuch fur Geologie und Paläontologie—Abhandlungen*, **249**, 119–27. A detailed theoretical treatment of extreme ocean acidification as an explanation for the end-Triassic coral gap and carbonate crisis in the seas is given in: Martindale, R.C. et al. (2012) Constraining carbonate chemistry at a potential ocean acidification event (the Triassic–Jurassic boundary) using the presence of corals and coral reefs in the fossil record. *Palaeogeography, Palaeoclimatology, Palaeoecology*, **350–352**, 114–23.
49. See Barnosky et al. (2011).
50. See Barnosky et al. (2011).
51. Ceballos, G. et al. (2015) Accelerating modern human-induced species losses: entering the sixth mass extinction. *Science Advances*, **1**, e1400253.
52. See Benton & Twitchett (2003).
53. D'Hondt, S. (2005) Consequences of the Cretaceous/Paleogene mass extinction for marine ecosystems. *Annual Review of Ecology, Evolution and Systematics*, **36**, 295–317.
54. Ehrlich, P.R. & Pringle, R.M. (2008) Where does biodiversity go from here? A grim business-as-usual forecast and a hopeful portfolio of partial solutions. *Proceedings of the National Academy of Sciences, USA*, **105 (Suppl 1)**, 11579–86.
55. See Butchart et al. (2010).
56. http:// www.cbd.int/sp/targets/
57. Joppa, L.N. et al. (2013) Achieving the Convention on Biological Diversity's goals for plant conservation. *Science*, **341**, 1100–3.
58. Tittensor, D.P. et al. (2014) A mid-term analysis of progress toward international biodiversity targets. *Science*, **346**, 241–4.
59. Zaccai, E. & Adams, W.M. (2012) How far are biodiversity loss and climate change similar as policy issues? *Environmental Development and Sustainability*, **14**, 557–71.
60. Gibson, D.G. et al. (2010) Creation of a bacterial cell controlled by a chemically synthesized genome. *Science*, **329**, 52–6.
61. Nee, S. & May, R.M. (1997) Extinction and the loss of evolutionary history. *Science*, **278**, 692–4. See also, Mace, G.M., Gittleman, J.L. & Purvis, A. (2003) Preserving the tree of life. *Science*, **300**, 1707–9.
62. See Mace et al. (2003).
63. Cowling, R.M. & Lombard, A.T. (2002) Heterogeneity, speciation/extinction history and climate: explaining region plant diversity patterns in the Cape Floristic Region. *Diversity and Distribution*, **8**, 163–79.
64. Davies, T.J. et al. (2011) Extinction risk and diversification are linked in plant biodiversity hotspot. *PLoS Biology*, **9(5)**, e1000620.
65. Forest, F. et al. (2007) Preserving the evolutionary potential of floras in biodiversity hotspots. *Nature*, **445**, 757–60.

66. Constanza, R. et al. (1997) The value of the world's ecosystem services and natural capital. *Nature*, **387**, 253–7.
67. Ring, I. et al. (2010) Challenges in framing the economics of ecosystems and biodiversity: the TEEB initiative. *Current Opinion in Environmental Sustainability*, **2**, 15–26.
68. Based on nearly 2 million records of species abundance, nearly 40 000 species from nearly 20 000 locations. See Newbold, T. et al. (2016) Has land use pushed terrestrial biodiversity beyond the planetary boundary? A global assessment. *Science*, **353**, 288–91. See also the commentary: Oliver, T.H. (2016) How much biodiversity loss is too much? *Science*, **353**, 220–1.
69. If those rivets represent common species, the analogy finds recent support here: Winfree, R. et al. (2015) Abundance of common species, not species richness, drives delivery of a real-world ecosystem service. *Ecology Letters*, **18**, 626–35.
70. http://www.footprintnetwork.org/
71. http://www.footprintnetwork.org/en/index.php/GFN/page/personal_footprint/
72. Kitzes, J. et al. (2008) Shrink and share: humanity's present and future ecological footprint. *Philosophical Transactions of the Royal Society*, **B363**, 467–75.
73. See Foley et al. (2011).
74. Sanchez, P.A. & Swaminathan, M.S. (2005) Cutting world hunger in half. *Science*, **307**, 357–9.
75. See Sanchez & Swaminathan (2005).
76. See Foley et al. (2011).
77. Ridley, M. (2010) *The Rational Optimist: How Prosperity Evolves*. Fourth Estate, London.
78. Trewavas, A. (2001) Urban myths of organic farming. *Nature*, **410**, 409–10.
79. Seufert, V., Ramankutty, N. & Foley, J.A. (2012) Comparing yields of organic and conventional agriculture. *Nature*, **485**, 229–32. See also the commentary by Reganold, J.P. & Doberman, A. (2012) Comparing apples with oranges. *Nature*, **485**, 176–7.
80. Cassman, K.G., Dobermann, A. & Walters, D.T. (2002) Agroecosystems, nitrogen-use efficiency, and nitrogen management. *Ambio*, **31**, 132–40.
81. Baulcombe, D. et al. (2009) *Reaping Benefits: Science and the Sustainable Intensification of Global Agriculture*. Royal Society Policy document, 11/09. The Royal Society, London.
82. Raven, P.H. (2010) Does the use of transgenic plants diminish or promote biodiversity? *New Biotechnology*, **27**, 528–33.
83. Tilman, D. & Clark, M. (2014) Global diets linked to environmental sustainability and human health. *Nature*, **515**, 518–22.
84. Stehfest, E. et al. (2009) Climate benefits of changing diet. *Climate Change*, **95**, 83–102.
85. Hansen, J. et al. (2008) Target atmospheric $CO_2$: where should humanity aim? *The Open Atmospheric Science Journal*, **2**, 217–31.
86. Hansen, J. et al. (2017) Young people's burden: requirement of negative $CO_2$ emissions. *Earth Systems Dynamics*, **8**, 577–616.
87. Clark, P.U. et al. (2016) Consequences of twenty-first-century policy for multi-millennial climate and sea-level change. *Nature Climate Change*, **6**, 320–69. See also Hansen et al. (2017).
88. Davidson, D.J. & Andrews, J. (2013) Not all about consumption. *Science*, **339**, 1286–7.

89. Edenhofer, O. et al. (eds) (2014) *Intergovernmental Panel on Climate Change, Climate Change 2014: Mitigation of Climate Change*. Cambridge University Press, New York.
90. See Le Quéré et al. (2016).
91. Baker, J.A. et al. (2017) *The Conservative Case for Carbon Dividends.* Climate Leadership Council. Report available at: https://www.clcouncil.org/media/TheConservativeCaseforCarbonDividends.pdf
92. https://citizensclimatelobby.org/
93. Quotes are from an interview with the author at the Missouri Botanic Garden (19 August 2013).
94. Raven offered me a final mischievous remark in this context: 'That reminds me of another thing someone once said: "You know, to solve all these environmental problems, nations of the world would have to come together and agree to work together, and we might need to form an organization; we might call it the United Nations".'
95. Sellars, S. & O'Hara, D. (2012) *Extreme Metaphors: Selected Interviews with J.G. Ballard, 1967–2008*. Fourth Estate, London.
96. See Ehrlich & Pringle (2008).

# FIGURE CREDITS

Figure 2 Chang, C., Bowman, J.L. & Meyerowitz, E.M., Field guide to plant model systems. *Cell*, 167, 325–339. © 2016 Elsevier Inc.

Figure 3 Reproduced and modified from Springer Nature: *Nature Plants, Enabling the Water to Land Transition*, Reski, R. © 2017 Macmillan Publishers Limited, part of Springer Nature. All rights reserved.

Figure 4 Reproduced from *Trends in Plant Science*, Vol.7, Hidalgo, O. et al., 'Is there an upper limit to genome size?', pp. 567–573. © 2017 with permission from Elsevier Ltd. All rights reserved.

Figure 5 Reprinted by permission from Springer Nature: *Nature*, 'Ancestral polyploidy in seed plants and angiosperms', Jiao, Y. et al. Copyright © 2011, Springer Nature.

Figure 6 Reproduced from *Current Opinion on Plant Biology*, Vol. 30, Soltis, P.S. & Soltis, D.E., 'Ancient WGD events as drivers of key innovations in angiosperms', pp. 159–165. © 2016 with permission from Elsevier Ltd. All rights reserved.

Figure 7 Courtesy of City of Hope Archives.

Figure 8 Reprinted by permission from Springer Nature: *Nature, Nature Plants*, 1, 'Expanding the role of botanical gardens in the future of food', Miller, A.J. et al. © 2015.

Figure 9 Reprinted by permission from Springer Nature: *Nature*, 'Stem cells that make stems', Weigel, D. & Jurgens, G., © 2002 Macmillan Publishers Limited, part of Springer Nature. All rights reserved.

Figure 10 Reprinted from 'The origin and early evolution of roots', Paul Kenrick, Christine Strullu-Derrien, *Plant Physiology* Oct 2014, 166 (2) 570–580; DOI: 10.1104/pp.114.244517. © 2014 American Society of Plant Biologists. All Rights Reserved.

Figure 11 Reproduced from *Current Biology*, Vol. 26, Arteaga-Vazquez, M.A., 'Land plant evolution: listen to your elders', R22-R40. © 2016 with permission from Elsevier Ltd. All rights reserved.

Figure 12 Reprinted by permission from Springer Nature: *Nature*, 'A vascular conducting strand in the early land plant Cooksonia', D. Edwards, K.L. Davies, L. Axe. © 1992, Springer Nature.

Figure 13 Redrawn from *Current Opinion in Plant Biology*, Vol. 13, Berry, J.A., Beerling, D.J. & Franks, P.J., 'Stomata: key players in the earth system, past and present', pp. 233–240. Copyright © 2010 with permission from Elsevier Ltd. All rights reserved.

Figure 16 Reproduced with permission from Cao, L., Bala, G., Caldeira, K., Nemani, E. & Ban-Weiss, G., 2010, 'Importance of carbon dioxide physiological forcing to future climate change', *Proceedings of the National Academy of Sciences*, USA, 107, 9513–9518.

Figure 17 (a) Jim Laws / Alamy Stock Photo. (b) Courtesy of Charles Wellman.

Figure 18 Based on Kidston R, Lang WH. (1921), On Old Red Sandstone plants showing structure, from the Rhynie Chert Bed, Aberdeenshire. Part IV. Restorations of the vascular cryptogams, and discussion on their bearing on the general morphology of the pteridophyta and the origin of the organization of land-plants', *Transactions of the Royal Society of Edinburgh*, 52, 831–854.

Figure 19 (a) Courtesy of Kris Pirozynski. (b) Courtesy of David Malloch.

Figure 20 David Beerling.

Figure 21 Leake, J.R., Cameron, D.D. & Beerling, D.J., 'Fungal fidelity in the myco-heterotroph-to-autotroph life cycle of Lycopodiaceae: a case of parental nurture', *New Phytologist*, 177, 572–576. Copyright © 2008, John Wiley and Sons.

Figure 22 Martin, F., Uroz, S. & Barker, D.G., 'Ancestral alliances: plant mutualistic symbioses with fungi and bacteria', *Science*, 356. Copyright © 2017, American Association for the Advancement of Science.

Figure 23 Courtesy New York State Museum, Albany, NY, USA.

Figure 24 Redrawn from Springer Nature: *Nature Communications*, G.L., Royer, D.L. & Lunt, D.J., 'Future climate forcing potentially without precedent in the last 420 million years,' doi:10.1038/ncomms14845. © 2017. Published by Springer Nature under the terms of the Creative Commons Attribution 4.0 International license. (https://creativecommons.org/licenses/by/4.0/).

Figure 25 Wellcome Collection. Reproduced under the terms of the Creative Commons Attribution 4.0 International license (https://creativecommons.org/licenses/by/4.0/).

Figure 26 Reproduced with permission by Springer Nature: *Nature*, 'Biodiversity hotspots for conservation priorities', Myers, N. et al. Copyright © 2000, Springer Nature.

Figure 27 Courtesy of Peter H. Raven.

# PLATE CREDITS

Plate 1 De Vries, J, & Archibald, J.M., 'Plant evolution: landmarks on the path to terrestrial life', *New Phytologist*, 217, 1428–1434, Wiley. © 2018 The Authors. New Phytologist © 2018 New Phytologist Trust.

Plate 2 Delwiche, C.F. & Cooper, E.D., The evolutionary origin of a land flora. *Current Biology*, 25, R899–R910. © 2015 Elsevier Inc.

Plate 3 Delwiche, C.F. & Timme, R.E., Plants. *Current Biology*, 21, R417–R422. © 2011 Elsevier Inc.

Plate 4 Reprinted by permission from Springer Nature: *Nature Plants*, 'Newton and the ascent of water in plants,' Beerling, D.J. © 2015 Macmillan Publishers Limited, part of Springer Nature. All rights reserved.

Plate 5 Reprinted by permission from Springer Nature: Nature, *Nature Plants*, 2, 'Origin and function of stomata in the moss *Physcomitrella patens*', Chater, C.C. et al. © 2016.

Plate 6 Remy, W. et al., 'Four hundred-million-year-old vesicular arbuscular mycorrhizae', *Proceedings of the National Academy of Sciences, USA*, 91, 11841–11843. © 1994, National Academy of Sciences, U.S.A. Courtesy of Hans Kerp (Münster, Germany).

Plate 7 David Beerling.

Plate 8 © 2012 Victor O. Leshyk.

Plate 9 David Beerling.

Plate 10 (a) David Beerling. (b) Morris, J.L. et al., Investigating Devonian trees as geo-engineers of past climates: linking palaeosols to palaeobotany and experimental geobiology. *Palaeontology*, 58, 787–801. © 2015 The Authors. Palaeontology published by John Wiley & Sons Ltd on behalf of The Palaeontological Association. Published by John Wiley & Sons, Inc. under the terms of the Creative Commons Attribution 4.0 International license. (https://creativecommonsorg/licenses/by/4.0/).

Timescale: Cohen, K.M., Finney, S., and Gibbard, P.L., 2012, International Chronostratigraphic Chart: International Commission on Stratigraphy, http:// www.stratigraphy.org (last accessed May 2012). (Chart reproduced for the 34th International Geological Congress, Brisbane, Australia, 5–10 August 2012.)

Gradstein, F.M, Ogg, J.G., Schmitz, M.D., et al., 2012, The Geologic Time Scale 2012: Boston, USA, Elsevier, DOI: 10.1016/B978-0-444-59425-9.00004-4.

# PUBLISHER'S ACKNOWLEDGEMENTS

We are grateful for permission to include the following copyright material in this book.

Extract from Lone Frank, *My Beautiful Genome*, OneWorld Publications, 2011. © 2010, Lone Frank. Reproduced with permission of the Licensor through PLSclear.

Extract from Winifred Goldring, 'The Oldest Known Petrified Forest', *The Scientific Monthly*, Vol. 24, No. 6 (Jun., 1927), pp. 514–29.

Extract republished with permission of Hachette Books Group, from *Symbiotic Planet: A New Look at Evolution*, Lynn Margulis, 1998; permission conveyed through Copyright Clearance Center, Inc.

Extract from *Lord of the Rings: The Return of the King*, reprinted by permission of HarperCollins Publishers Ltd. © 1955 J.R.R. Tolkien

Extract reprinted from *Current Biology*, Vol. 23, Issue 21, Keiko Torii, 'Q & A with Keiko U. Torii', R943–4. © 2013, with permission from Elsevier.

Extract reprinted by permission from Springer Nature, *Nature*, Multiple personal genomes await, J. Craig Venter, 2010. © 2018 Macmillan Publishers Limited, part of Springer Nature. All rights reserved.

The publisher and author have made every effort to trace and contact all copyright holders before publication. If notified, the publisher will be pleased to rectify any errors or omissions at the earliest opportunity.

# INDEX